# GLOBALIZATION, FOREIGN DIRECT INVESTMENT AND TECHNOLOGY TRANSFERS

The impact of globalization during the past decade has been uneven across regions, countries and sectors. While many countries have become more integrated into the world economy, increasing their cross-border transactions, others have been left behind. Foreign direct investment (FDI) inflows and technology transfers are increasingly concentrated in a handful of countries, while the least developed countries continue to be marginalized, despite the liberalization of their investment regimes. *Globalization, Foreign Direct Investment and Technology Transfers* discusses emerging patterns and directions of FDI and technology transfers and the growth prospects for developing countries in this context of globalization.

In this study, leading scholars on the subject examine and explain the emerging patterns in international technology transfers and FDI flows over the past two decades. They analyse the trends in internationalization of corporate activity in major conventional and emerging source countries of FDI to comment upon the prospects for outflows. This departs from the existing treatments of FDI as a homogeneous resource and allows for a more detailed prediction of future outflow patterns. Throughout, the research focuses upon the implications of new trends for developing countries. Kumar concludes by outlining the policy implications for the governments of such countries seeking to mobilize technology and FDI for their industrialization and further integration into the international community. Controversially, he cautions against excessive optimism about the potential of FDI inflows as an agent of development.

*Globalization, Foreign Direct Investment and Technology Transfers* draws together much data and information which is not readily available and provides reflections upon international business negotiations from a developing country's perspective.

**Nagesh Kumar** is currently Senior Fellow at the Research and Information System for Developing Countries, New Delhi, India. From 1993 to 1998, Dr Kumar served on the faculty of the United Nations University Institute for New Technologies, Maastricht, the Netherlands and directed its research on foreign direct investment and technology transfers.

# UNU/INTECH STUDIES IN NEW TECHNOLOGY AND DEVELOPMENT

Series editors: Charles Cooper and Swasti Mitter

The books in this series reflect the research initiatives at the United Nations University Institute for New Technologies (UNU/INTECH) based in Maastricht, the Netherlands. The institute is primarily a research centre within the UN system and evaluates the social, political and economic environment in which new technologies are adopted and adapted in the developing world. The books in the series explore the role that technology policies can play in bridging the economic gap between nations, as well as between groups within nations. The authors and contributors are leading scholars in the field of technology and development; their work focuses on:

- the social and economic implications of new technologies;
- processes of diffusion of such technologies to the developing world;
- the impact of such technologies on income, employment and environment;
- the political dynamics of technological transfer.

The series is a pioneering attempt at placing technology policies at the heart of national and international strategies for development. This is likely to prove crucial in the globalized market, for the competitiveness and sustainable growth of poorer nations.

# GLOBALIZATION, FOREIGN DIRECT INVESTMENT AND TECHNOLOGY TRANSFERS

Impacts on and prospects for developing countries

*Nagesh Kumar*

in collaboration with
*John H. Dunning, Robert E. Lipsey,
Jamuna P. Agarwal and Shujiro Urata*

London and New York

The United Nations
University

**INTECH**

Institute for New Technologies

Published in association with the UNU Press

First published 1998
by Routledge
11 New Fetter Lane, London EC4P 4EE

Simultaneously published in the USA and Canada
by Routledge
29 West 35th Street, New York, NY 10001
© 1998 UNU/INTECH

Typeset in Garamond by
Pure Tech India Ltd., Pondicherry
Printed and bound in Great Britain by
Biddles Ltd, Guildford and King's Lynn

*British Library Cataloguing in Publication Data*
A catalogue record for this book is available from the British Library

*Library of Congress Cataloging in Publication Data*
A catalogue record for this book has been requested

ISBN 0–415–19111–4

# CONTENTS

CONTENTS

# FIGURES

xi

# TABLES

# APPENDICES

# CONTRIBUTORS

**Dr Jamuna P. Agarwal** is a Senior Economist at the Kiel Institute of World Economics in Kiel, Germany.

**Professor John H. Dunning** has held chairs in international business at the University of Reading, UK, and at Rutgers University, USA. He has also been associated with UNCTAD as the Senior Economic Adviser for its Programme on Transnational Corporations.

**Dr Nagesh Kumar** is currently Senior Fellow at the Research and Information System for Developing Countries, New Delhi, India. From 1993 to 1998, Dr Kumar served on the faculty of the United Nations University Institute for New Technologies, Maastricht, the Netherlands. He received the Exim Bank of India's first International Trade Research Award for his book *Multinational Enterprises in India: Industrial Distribution, Characteristics and Performance*, also published by Routledge in 1990.

**Professor Robert E. Lipsey** is Director, New York Office of the National Bureau of Economic Research. He is also a Professor Emeritus of Queens College and the Graduate Centre of the City University of New York.

**Professor Shujiro Urata** is Professor of Economics, Waseda University, Tokyo, Japan. He is also affiliated to the Japan Centre for Economic Research.

# PREFACE

The contributions included in this book have been prepared as a part of the UNU/INTECH research project on foreign direct investment, technology transfers and export-orientation in developing countries. They are intended to provide a background to a quantitative analysis of the factors that shape the role that MNEs play in industrialization, technology transfer, export expansion and innovative activity of their host countries. The quantitative studies themselves are reported in a forthcoming companion volume entitled *Globalization and the Quality of Foreign Direct Investment Inflows: A Quantitative Explanation of the Role of Multinationals in Industrialization, Export-expansion and Innovation in the Host Countries*. A selective review of existing literature on some of the issues addressed in this volume has been presented in parts of *Technology, Market Structure and Internationalization: Issues and Policies for Developing Countries* (Nagesh Kumar and N. S. Siddharthan, Routledge and UNU Press, 1997).

The completion of this book gives me an opportunity to acknowledge the intellectual debt accumulated in the course of its preparation. My foremost gratitude is to Professor Charles Cooper, UNU/INTECH's director, who has lent his wholehearted support to the project and has offered valuable advice whenever sought. I would like to thank my collaborators in this volume for agreeing to contribute their inputs and cooperation by meeting different deadlines. Most of these contributions were discussed in their draft form at an international workshop held in Maastricht in November 1996. We are grateful to the participants for their comments, especially to Professor Danny Van Den Bulcke of the University of Antwerp – RUCA, and Dr Kwang Jun of the World Bank for their discussions of the papers presented in this volume. Professor Van Den Bulcke also read the whole manuscript thoroughly and made many valuable suggestions.

A number of colleagues and friends have contributed to the project in different stages through discussions and by sharing data and information. I benefited from the discussions with Larry Westphal of Swarthmore College in the United States and formerly of UNU/INTECH in the initial stages of the project. Rakesh Mohan of the National Council of Applied Economic

Research, New Delhi, who visited UNU/INTECH during 1993/4, also provided helpful discussions on many themes concerning FDI and technology transfer. Sanjaya Lall of Oxford University commented on an earlier version of Chapter 2. Laudeline Auriol and François Pham of OECD/DSTI introduced me to various national sources of technology transfer data in OECD countries which proved instrumental for the analysis of Chapter 2. The late Professor Shoji Ito of Yokohama City University shared with me some data on technology exports by Japanese enterprises. I feel sorry that he is not to see the use to which these data have been put. Roger van Hoesel of Erasmus University Rotterdam generously shared some sources on FDI and technology transfers of Korean and Taiwanese companies used in Chapters 2 and 7. My academic colleagues at UNU/INTECH have supported the project with enthusiasm. Swasti Mitter has been a source of encouragement and advice throughout. Ludovico Alcorta has shared thoughts and advice through different stages of the project implementation over numerous luncheon discussions.

I must sincerely thank Alexandra van der Poel for her cheerful handling of all secretarial tasks relating to the project, organization of the November 1996 Workshop, and preparation of the present manuscript. Jane Williams has coordinated with the Routledge editorial team. But for her enthusiasm the completion of this manuscript would have taken considerably longer.

Finally a more personal note; I should like to record my sincere appreciation of Rachana for her patient bearing with my long hours of work and for providing moral support. It is significant that this manuscript should be completed and signed today, the day we were married. Hence, it comes as an anniversary present to her!

Nagesh Kumar
Maastricht, the Netherlands
2 December 1997

# 1

# INTRODUCTION

*Nagesh Kumar*

## The context

One of the most notable trends in the world economy over the past decade has been its increasing global economic integration. The world economy has witnessed a growing internationalization reflected in terms of the rising share of international trade and foreign direct investment (FDI) flows. The average annual growth rate of world merchandise trade has been twice as much as that of world output during the second half of the 1980s, and over three times during the first half of the 1990s. FDI inflows have expanded at an annual rate of 24 per cent during the second half of the 1980s and at 17 per cent during the 1991–6 period to touch a peak of \$349 billion in 1996. Another development has been in the expansion of the world trade in commercial services which has grown at an average annual rate of 22 per cent during the second half of the 1980s. Financial services have been liberalized on a large scale since the late 1980s following the deregulation of financial markets worldwide and the evolution of new information technologies. A number of business services such as banking, insurance, advertising, accounting, communications, media, car rental and catering services have also become increasingly internationalized over the past decade. The internationalization of these services, especially media, advertising and communication services, is highly visible. It affects consumption and demand patterns across the world and gives a greater visibility and edge to multinational enterprises. A cumulative effect of these trends has been to make the world economy far more integrated economically in the 1990s than ever before.

The impact of globalization has been uneven across regions, countries and sectors. Some countries and regions have integrated with the world economy more deeply with growing magnitudes of inward and outward FDI flows and other types of cross-border transactions while others have been left behind. New sources of FDI and technology have come up. Japan has emerged as one of the most important sources of FDI in this period as Japanese corporations were forced to move production abroad in order to stay competitive in the face of rising wages and an appreciating yen. FDI outflows from newly industrializing economies (NIEs), which had been noticed in the late 1970s and early 1980s, became an established trend by the late 1980s when several enterprises from South Korea and Taiwan, among others, grew multinational in their own right. The concentration of FDI inflows in a handful of countries has intensified with China emerging as one of the most important hosts of FDI inflows in the world over the past decade. The least developed countries, on the other hand, continue to receive a negligible share of FDI inflows despite liberalization of their investment regimes. The distribution of technology flows is also characterized by a similar concentration in a few countries.

The geography of FDI and technology transfers originating in different source countries has changed over the years in a different manner. These changes have been caused largely by changing policies of home and host developing countries, their developmental patterns and their growth prospects, but in no small measure by a changing external environment. The latter includes the global trend of liberalization of national economies to trade and investments, privatization of public sector enterprises and opening of services and infrastructure sectors, the trend of regional economic integration in several regions, the emergence of new generic or core technologies, economic reforms in Eastern and Central European countries and rapid industrialization and technological learning in the East Asian countries, among others. These developments have had wide-ranging implications for the organization of economic activity in the world economy. For instance, the evolution of new core technologies, e.g. ICT (information and communication technologies), biotechnologies and advanced materials and their widespread application in different sectors highlighted their commercial potential and have prompted a reaction of technonationalism and technological protectionism in the industrialized world. This has led to a wave of corporate restructuring in the Western world reflected into a sudden spurt of strategic alliances among corporations besides mergers and acquisitions. The deep economic integration in Europe and North America has also sparked off worldwide corporate restructuring as corporations attempted to restructure their operations on a pan-regional basis to gain efficiency and indulged in acquisitions of enterprises with complementary assets and refocused on their core competencies. The institutional framework for technology generation changed with the adoption of the Agreement on Trade Related Intellectual Property Rights (TRIPs) as a part of the multilateral trade negotiations providing for the harmonization and strengthening of the intellectual property system worldwide. This has made the imitation based technology acquisition strategy, widely practised by the East Asian countries, more difficult.

A number of international and national publications regularly report and discuss short-term trends in the annual flows of FDI flows (see, for instance, UNCTAD's series of annual *World Investment Reports*; OECD's *International Direct Investments Statistics Yearbook*; World Bank's *World Debt Tables and*

*Global Economic Prospects and the Developing Countries*; US Department of Commerce's *Survey of Current Business*; *Monthly Report of Deutsche Bundesbank*; UK's *Business Monitor* MA4; JETRO's Annual *White Paper on Foreign Direct Investment*, among others). However, a comprehensive perspective on the direction and patterns in FDI outflows and technology transfers emerging through their period of metamorphosis over the past two decades has been lacking, especially from the point of view of developing host countries. This book makes a modest attempt in that direction.

The contributions in this book examine the emerging trends and patterns in the international technology transfers and foreign direct investments flows over the past two decades, suggest their explanations and analyse their implications for developing countries. The trends of internationalization of corporate activity in a few major source countries of FDI and its impact on the developing host countries is then studied in greater detail. The book also examines the recent trends and patterns in FDI flows emanating from newly industrializing countries in Asia which are becoming important sources of FDI for developing countries. The volume concludes with policy implications for developing countries seeking to mobilize technology and FDI for their industrialization and development.

The rest of this chapter is devoted to a brief review of the contemporary theory of international operations of firms that treats FDI and arm's length licensing as alternative channels of the international transfer of disembodied technology. This discussion provides a conceptual backdrop to some of the explanations for the emerging trends offered in this volume in terms of the terminology of the theory. Finally, a chapter scheme is presented to provide an overview of the contents of the book.

## FDI as a channel of technology transfer – the theoretical treatment

The theory of international operations of firms as evolved over time with the contributions of Hymer (1960), Caves (1971), Buckley and Casson (1976) and Dunning (1981), among others, considers FDI and arm's length contracts as alternative modes of foreign production and technology transfer. In the theory, a firm wanting to operate abroad must possess advantages that are sufficient to more than offset the handicaps involved in functioning in an alien atmosphere and in covering the risks that are inherent in such operations. These advantages emanate from ownership (hence called ownership or 'O' advantages) of some proprietary intangible assets by firms such as product and process technology (patented or otherwise), brand ownership, managerial and marketing skills and access to cheaper sources of capital and raw materials. In the early phases of the 'product cycle' these advantages are exploited abroad through exports from the country of origin (Vernon, 1966). In the later phases of the product when competition builds up, foreign production is undertaken in the countries offering locational or 'L' advantages. These advantages arise from factors such as communication and transport costs, inter-country differences in input/factor prices and productivity. The foreign production is undertaken through FDI or within the firm (internally) if the market transactions are difficult to set up and govern. In other cases external market contracts are used to license the intangible assets. Thus the choice between FDI and licensing is determined by the transaction or governance costs. The higher the transaction costs, the higher the incentive to internalize the transaction (internalization or 'I' incentives) and likelihood of FDI being chosen as a mode of foreign production or technology transfer. The transaction costs are generally high for market transactions because of market failures due to 'public good like' nature of technology, difficulties in making a convincing disclosure and buyer's uncertainty, problems with codification of knowledge, and risks of dissipation of brand goodwill. It must be pointed out, however, that external markets for all intangible assets are not subject to the same degree of market failure, and hence transaction costs vary. Some intangible assets such as proprietary process technologies can be profitably licensed at arm's length. Product technologies and those process technologies that cannot be codified easily or embodied in capital goods because of a high tacit component are more difficult to license. Therefore, the relative importance of licensing as a channel of technology or knowledge transfers varies across industries.

Besides the characteristics of intangible assets or technology to be transferred as predicted by theory, a number of other factors may affect the choice between FDI and licensing in practice. For instance, licensing is preferred when FDI is not profitable or possible. This could be because of the small size of the market or government restrictions on FDI. Licensing is encouraged when the licenser lacks experience in managing manufacturing plants abroad (Caves, 1996). Davidson and McFetridge (1985) in an empirical study of the choice of mode of foreign participation found certain firm- and technology-specific factors also playing a role such as age, size, past experience in technology transfer, extent of product and geographical diversification of the firm and the degree of sophistication of technological advance, among others. An OECD survey of licensing has shown that smaller enterprises engaged in licensing activity unaccompanied by equity ownership more than larger corporations (Vickery, 1988). Local absorptive capacity and entrepreneurship are also important for licensing as the survey showed that most licensing deals arose from requests from host enterprises and that they were concentrated in industrialized countries and in the developing countries in Asia. Process technologies were found to be transferred on a licensing basis more often than product technologies. Finally, most licensing agreements included restrictive clauses such as territorial limits, exclusivity provisions and market limitations. Inter-industry analyses have also observed the different patterns of FDI and licensing across industries. Kumar (1987), in an analysis of determinants of FDI and licensing across 49 Indian manufacturing industries, found FDI to predominate in the advertising and human skill intensive industries. Licensing was important in industries where knowledge could be embodied in capital goods and those with relatively simpler technologies. Kumar (1990) found the US MNEs preferring FDI for transfer of product technologies, and licensing for transferring process technologies.

In the period since the mid-1960s, a considerable proportion of international transfer of disembodied technology has taken place under arm's length contracts or licensing arrangements (Dunning, 1993). This was because maturing and standardization of a wide range of technologies led to widening of technology markets as alternative sources emerged and competition increased. A large number of host governments evolved foreign investment codes during the 1960s and 1970s and started restricting FDI inflows in an effort to reduce remittances of dividends and/or to protect domestic enterprises. Thus arm's length licensing emerged as an alternative channel of international technology transfer. This trend of the rising importance of arm's length contracts as an alternative to FDI continued till the mid-1980s. As we later observe in this book, this trend has been reversed since the mid-1980s and FDI has regained some of its lost importance as a mode of technology transfer. This reversal of trend has been explained in terms of some of the global developments described earlier such as evolution of generic core technologies, changing attitudes of host governments towards FDI and wide-ranging liberalization of national economies.

## This book: chapter layout

The subject matter of the book has been classified into four parts. Part I deals with the changing geography of international technology transfer and FDI over the past two decades, their explanations and implications. Specifically, Kumar in Chapter 2 examines the trends and patterns in technology generation and international technology transfers in the world economy over the 1975–95 period with the help of data compiled from different national and international sources. In particular he examines the trends in the country and corporate concentration in the technology generation in terms of indicators of R&D activity, patent ownership and receipts of technology fees and royalties against the background of emergence of new core technologies and the recent trend of corporate restructuring. The chapter then examines trends and patterns in international technology transfers as represented by royalty and technology receipts by major source countries of technology and contrasts these trends with those in FDI inflows. The trends with respect to the mode of technology transfer are then discussed. It concludes with a few observations for technology import policies.

Against the background of the fact that FDI is a major channel of international technology transfer, Dunning in Chapter 3 identifies the changes that have taken place in the geography of FDI inflows since the mid-1970s and suggests reasons for them. These changes, which are the most profound of those that have taken place over the past two centuries, both affect and are affected by the geography of production. He also examines some of the implications of the changing geography of FDI for the cross-border transfer of knowledge capital and other created assets, and traces some of the consequences of the changing motivations for, and content of, MNE activity for the balance between inbound and outbound FDI, and for the interaction between these and domestic investment. The OLI theory provides the analytical framework for much of the analysis. The chapter then offers policy implications.

Part II presents more detailed case studies of FDI and technology outflows originating in conventional source countries of FDI and technology – the triad nations. As pointed out by Dunning in Chapter 3, the geography of FDI and hence of technology transfer is significantly affected by the country composition of outflows; a more detailed look at the patterns emerging in outflows from triad countries would be useful. We pick up the US, Germany and Japan for detailed examination. The bulk of the European FDI outflows are absorbed within the other regional member states of the European Union and a considerable and growing proportion of the outflows is directed to East and Central European countries. Germany's case is especially interesting from that point of view because German FDI outflows have shown a more pronounced eastwards shift that has been portrayed to be to the disadvantage of their investments in developing countries. Hence, Germany's case is picked up for exploration from among the European major outward investors to examine the implications of the geographical shift in German MNEs' focus. Japanese FDI outflows also reveal a strong regional focus: hence, special attention is paid to their investments in Southeast and East Asian developing countries.

Lipsey, in Chapter 4, examines the trends in the degree of internationalization of the US corporations in terms of internationalized production and its importance for the host countries – especially developing countries over the 1977–94 period. Then the chapter examines these trends with respect to the degree of self sufficiency of US affiliates to discover trends with respect to the creation of vertical inter-firm linkages by the US affiliates within the host economies. It then looks into the industry composition of US affiliates' production and discusses the issues relating to other impacts of overseas affiliates on host countries. Finally, the findings and their implications are summarized.

Chapter 5, by Agarwal, examines the recent growth of German FDI outflows and the patterns of its regional distribution. Noting that Central and East European countries have emerged as the important recipients of German FDI, the chapter analyses whether these countries are diverting FDI flows away from developing countries. To this end it discusses the main causes of relatively high involvement of German MNEs in Central and East European countries and motivations for their investments. It then examines the prospect of diversion of German investments away from developing countries in the future.

Chapter 6, by Urata, deals with Japanese FDI which is fast emerging as one of the most important sources of these flows. In particular, it examines the impact of Japanese FDI on developing Asian countries which have been major hosts of these flows among developing countries. It first examines the recent trends in Japanese FDI with an emphasis on Asia and examines their motivations. Then it analyses the impact of Japanese FDI in exports expansion and technology acquisition of Asian developing countries. It is concluded with some remarks on future prospects and policies.

Part III covers trends in the emerging sources of FDI and technology, their relative importance for the developing countries as sources of FDI and the prospects for developing countries. Kumar in Chapter 7 examines the trends and patterns in FDI outflows from the newly industrializing economies, especially those in Asia which have become significant sources of FDI in the past ten years. He argues that the recent spurt in the annual outflows of FDI from these countries has coincided with a change in their motivation from a market defensive type to more competitiveness seeking. The multinationals emerging from developing countries are important sources of FDI inflows, especially for developing countries, and their importance as sources of FDI is particularly striking for ASEAN and some Latin American countries. He also examines the implications of this trend for developing countries as potential hosts of FDI and concludes with a few remarks on policy.

Finally, Chapter 8 in Part IV pulls together the threads from the previous three parts on the prospects and implications, for developing countries, of the increasing globalization with expansion of FDI and other cross border activity. It examines prospects of further expansion of the scale of FDI inflows to developing countries. It discusses the role of factors that shape the distribution of FDI among the developing countries and draws out the implications for policies of the least developed countries. Then we summarize certain aspects of strategic interventions by developing country governments to maximize the gains from globalization. Finally, a few important issues of international action in the area of international business are discussed.

# Part I

# GEOGRAPHY OF INTERNATIONAL TECHNOLOGY TRANSFER AND FDI

This part deals with the changing geography of international technology transfer and FDI over the past two decades, their explanations and implications. Specifically, Chapter 2 examines the trends and patterns in technology generation and international technology transfers in the world economy over the period 1975–95 with the help of data compiled from different national and international sources. First the trends in the country and corporate concentration in technology generation are examined in terms of different indicators of R&D activity. Then trends and patterns in international transfers of disembodied technology by major source countries of technology are analysed and contrasted with those in FDI inflows. The trends with respect to the mode of technology transfer are then discussed. Against the background of the fact that FDI is a major channel of international technology transfer, Chapter 3 identifies the profound changes that have taken place in the geography of FDI inflows since the mid-1970s and suggests reasons for them. It also examines some implications of the changing geography of FDI for the cross-border transfer of knowledge capital and other created assets, and traces consequences of the changing motivations for the balance between inbound and outbound FDI and for their interaction with domestic investment.

# 2

# TECHNOLOGY GENERATION AND TRANSFERS IN THE WORLD ECONOMY

## Recent trends and prospects for developing countries

*Nagesh Kumar*

## Introduction

Technology is a crucial input in the industrialization and development of countries. In the current era of liberalization of international trade regimes worldwide and the increasing emphasis attached to international competitiveness, the importance of technology as a factor determining the growth prospects of nations has risen even further. A considerable proportion of technological inputs is sourced from abroad especially in the early stages of a country's development. Technology is perceived as an intangible asset and is traded internationally either in embodied or in disembodied form. The former includes the transfer of knowledge incorporated in the designs of machinery or that embodied in the skills of migrating experts. The disembodied knowledge is transferred under contracts under which process knowhow, product designs, rights to use patented knowledge or copyrighted designs or drawings are transferred by their owner to another party for a fee. The international technology markets continue to be dominated by a handful of industrialized countries. The technological activity within these countries is also known to be highly concentrated among a smaller set of bigger corporations dominating different branches of industries.

A number of changes have taken place over the past two decades or so which might have important implications for the patterns of technology generation and its transfer abroad. These developments include evolution of new core technologies, e.g. ICT (information and communication technologies), biotechnologies and advanced materials, and their widespread application in different industries has highlighted their commercial potential and prompted

11

a wave of technonationalism and technological protectionism in the industrialized world. The institutional framework for technology generation changed with the adoption of the Agreement on Trade Related Intellectual Property Rights (TRIPs) as a part of the Uruguay Round of multilateral trade negotiations providing for the harmonization and strengthening of the intellectual property system. The deep economic integration in Europe and North America has sparked off worldwide corporate restructuring. Finally, internationalization of R&D activity by multinational enterprises (MNEs) became an established trend. All these developments are likely to have implications for the international technology markets and hence technology transfers.

In the late 1970s and early 1980s a large volume of literature had addressed itself to the examination and analysis of trends and patterns in international technology transfer activity and their implications (see, for instance, Stewart, 1979; Lall, 1980; Madeuf, 1984; OECD, 1982; Rosenberg and Frischtak, 1985; Vickery, 1986, 1988; and Reddy and Zhao, 1990 for a review). However, in the more recent period, relatively little information and analysis on the emerging trends in international technology transfers is available despite the increasing importance of cross-border transactions in technology and their role in development. An examination of trends, and analysis of their implications for technology acquisition could prove to be valuable for policy making in developing countries which substantially depend on imports for fulfilling the technological requirements in the process of their industrialization.

This chapter pieces together information and data from diverse national and international sources to generate a rough picture of emerging trends in the global technology market and in international transfers of disembodied technology. We use technological payments data to capture disembodied technology transfers, assuming that any significant international transfers of knowledge are paid for. These data are contrasted with those on FDI inflows. It helps to examine the trends and patterns emerging in these transfers. These cover trends in concentration of sources of technology as well as patterns of distribution, and the relative importance of internal and external channels of transfer, among others. It would appear from this analysis that the triad nations consisting of the USA, the European Economic Area and Japan account for the bulk of origin as well as destination of technology flows. The importance of FDI or internal sources of technology transfer as a channel of technology transfer seems to have risen in recent years. The global flows of technology fees and global magnitudes of FDI inflows show a great degree of correspondence. Yet it becomes apparent that it may be hazardous to use inflow of FDI as an indicator of technology transfer. This chapter then concludes with the implications of the findings.

The structure of the chapter is as follows: the next section discusses trends in technology generation especially in country and corporate concentration; the third section examines trends in international transfers of disembodied technology and contrasts the patterns observed with those in FDI inflows; the fourth section looks at trends and patterns in the mode of technology transfers over time, across sectors and countries; and finally, the chapter is concluded with a few policy remarks.

## Trends in technology generation

The innovative activity which is the source of technology generation has been highly concentrated in a handful of industrialized countries. Besides being concentrated in a few countries, the innovative activity is dominated by a small number of larger corporations. In what follows, the trends pertaining to country as well as corporate concentration in technology generation are examined.

### *Country concentration in technology generation*

Table 2.1 compiles a few indicators of technological activity for the ten largest sources of technologies in the world. These indicators cover indicators of both technological 'inputs' as well as 'output'. R&D expenditure is considered as the most important input indicator of the technological activity. One of the technology output indicators considered is patents obtained by inventors from different countries at the US Patents and Trademarks Office over the past twenty-year period (1977–96). Since inventors from anywhere in the world like to register a patent in the US, the world's biggest market and one providing the most stringent and longest duration of patents (twenty years), US patents are widely considered as surrogates of technological output especially for the purposes of international comparisons. However, it is important to keep in mind that this measure may be biased in favour of industrialized countries, as much of the innovative activity in developing countries is of adaptive type and may not yield patentable inventions. Second, the considerable administrative costs of obtaining a US patent may discourage some inventors, especially those in developing countries, from applying for them. Another indicator of technological activity is the actual receipts of royalties and technological fees which indicate the amount of disembodied technology exported and hence measure the importance of a nation as a supplier of technology. The final column summarizes the figures on FDI outflows which are governed to a large extent by the strength of created assets of their enterprises – with ownership of technology being the most important source of the firms' created assets.

The data summarized in Table 2.1 reveal an extreme form of concentration with just ten countries accounting for the bulk of all technological activity in the world. It is evident that the top ten countries account for as much as 84 per cent of global resources spent on R&D activity annually; they control 95 per cent of the technological output in terms of patents taken

*Table 2.1* Major source countries of technologies, mid-1990s

| Country | R&D expenditure, 1993[1] | | US patents taken, 1977–96[2] | | Technology fees received, 1993[3] | | FDI outflows, 1995[4] | |
|---|---|---|---|---|---|---|---|---|
| | (billion ppp $) | % of total | '000 | % of total | billion $ | % of total | billion $ | % of total |
| USA | 166.3 | 39 | 985.3 | 57 | 20.4 | 40 | 95.5 | 30 |
| Japan | 74.4 | 17 | 307.6 | 18 | 3.6 | 7 | 21.3 | 7 |
| Germany | 37.1 | 9 | 136.2 | 8 | 7.3 | 14 | 35.3 | 11 |
| France | 26.4 | 6 | 52.7 | 3 | 2 | 4 | 17.5 | 6 |
| UK | 21.6 | 5 | 52.8 | 3 | 2.9 | 6 | 37.8 | 12 |
| Italy | 13.2 | 3 | 22.1 | 1 | 0.9 | 2 | 5.1 | 2 |
| Canada | 8.4 | 2 | 34.4 | 2 | 0.9 | 2 | 4.8 | 2 |
| Netherlands | 5.1 | 1 | 16.9 | 1 | 6.2 | 12 | 12.4 | 4 |
| Sweden | 4.8 | 1 | 17.3 | 1 | 0.4 | 1 | 10.4 | 3 |
| Switzerland | 4.2 | 1 | 25.5 | 1 | 2[6] | 4 | 8.6 | 3 |
| Subtotal 10 | 361.5 | 84 | 1,650.8 | 95 | 46.6 | 91 | 248.7 | 79 |
| World | 428.58[5] | 100 | 1,732 | 100 | 51[7] | 100 | 315 | 100 |

*Source*: Kumar, based on:
1 OECD (1996) *OECD in Figures – Statistics on the Member Countries: 1996 Edition*, Paris: OECD, pp. 56–7;
2 US Patents and Trademarks Office (1997) *TAF Special Report: All Patents, All Types – January 1977–December 1996*, Washington, DC;
3 OECD (1996), pp. 60–1;
4 UNCTAD (1996) *World Investment Report 1996*, Geneva: United Nations;
5 UNESCO (1996) *World Science Report 1996*, Paris: UNESCO. This figure relates to 1992;
6 own estimates based on mirroring of payments by major OECD countries; and
7 own estimate providing for non-reporting countries.

out in the US, and receive 91 per cent of global cross-border royalties and technological fees. Hence, the concentration in terms of technological output is even more uneven than for technological inputs. The control over technology is reflected in their nearly 80 per cent share of global FDI outflows.

The relative position of countries in terms of different indicators varies. It is curious to find, for instance, that Japan accounts for 17–18 per cent of global R&D expenditure and US patents, but shares only 7 per cent of global technological receipts (and FDI outflows). Similarly, the Netherlands and Switzerland command a much higher proportion of global royalties than their share of R&D expenditure or patents. The discrepancy between the share in technological activity and returns in terms of royalties could be on account of the different focus of the technological effort of different countries. The technological effort of Japanese corporations seems to be geared towards making their products competitive in domestic and export markets more than towards supporting the transfer of technology and production. Audretsch and

Yamawaki (1988) made a similar inference from their detailed empirical study of R&D rivalry and trade balance between the US and Japan.

Has the concentration in generation or ownership of technology declined over time? An over time examination of trends is only possible in terms of US patents ownership and royalty receipts, as reliable data on R&D expenditure on a long term comparable basis in constant prices is difficult to find. Table 2.2 summarizes the pattern of ownership of US patents over the 1977–96 period. It would appear from it that although the share of patents owned by US inventors has gone down over time, the share of the top ten patent-owning countries has not declined much. Japan has emerged strongly as a leading holder of US patents improving its share steadily from 11 per cent in pre-1983 patents to nearly 21 per cent in the 1990s. The share of the top ten countries in total patents has declined slightly from 96.2 per cent during the 1977–82 period to 96.0 during 1983–9 and to 94.4 per cent during the 1990–6 period, suggesting a slight decline in the overall concentration. However, there is a slight tendency of a higher concentration among the top three innovating countries – the US, Japan and Germany – with their combined share rising from 82 per cent in the first period to 83 per cent in the latest period. The trends in terms of receipts of royalty and technological fees reveal more changes. Figure 2.1 summarizes the composition of the receipts of royalties received by major OECD countries (excluding Switzerland for lack of data) in 1976 and 1993. It also suggests the decline in the relative position of the US as a source of technology with its share going down from 53 to 43

*Table 2.2*  Trends in ownership of US patents, 1977–96

| Country | Patents granted during the period, and percentage | | | | | |
|---|---|---|---|---|---|---|
| | *1977–82* | *%* | *1983–9* | *%* | *1990–6* | *%* |
| US | 244,507 | 62.1 | 556,267 | 58.0 | 429,052 | 55.5 |
| Japan | 43,977 | 11.2 | 147,441 | 15.4 | 160,167 | 20.7 |
| Germany | 34,237 | 8.7 | 84,545 | 8.8 | 51,625 | 6.7 |
| United Kingdom | 15,002 | 3.8 | 33,753 | 3.5 | 19,043 | 2.5 |
| France | 12,551 | 3.2 | 30,959 | 3.2 | 21,790 | 2.8 |
| Canada | 7,223 | 1.8 | 18,089 | 1.9 | 16,305 | 2.1 |
| Switzerland | 7,581 | 1.9 | 16,564 | 1.7 | 8,910 | 1.2 |
| Italy | 4,757 | 1.2 | 12,311 | 1.3 | 9,773 | 1.3 |
| Sweden | 5,035 | 1.3 | 11,455 | 1.2 | 5,870 | 0.8 |
| Netherlands | 3,955 | 1.0 | 10,072 | 1.0 | 6,817 | 0.9 |
| Share top 10 | 378,825 | 96.2 | 921,456 | 96.0 | 729,352 | 94.4 |
| Share top 3 | 322,721 | 82.0 | 788,253 | 82.2 | 640,844 | 82.9 |
| Total | 393,629 | 100.0 | 959,368 | 100.0 | 772,927 | 100.0 |

*Source*:  Kumar, based on data presented in US Patents and Trademarks Office (1997) *TAF Special Report: All Patents, All Types – January 1977–December 1996*, Washington, DC.

per cent between 1976 and 1993. Another feature apparent from the figure is the decline of the UK as the second most important source of technology in 1976, with a 23 per cent share, to fifth place in 1993, accompanied by the emergence of Germany, Japan and the Netherlands as major sources of techno- logy. On the whole the concentration would appear to have loosened a bit with the top five sources accounting for 84 per cent share in 1993 in place of 89 per

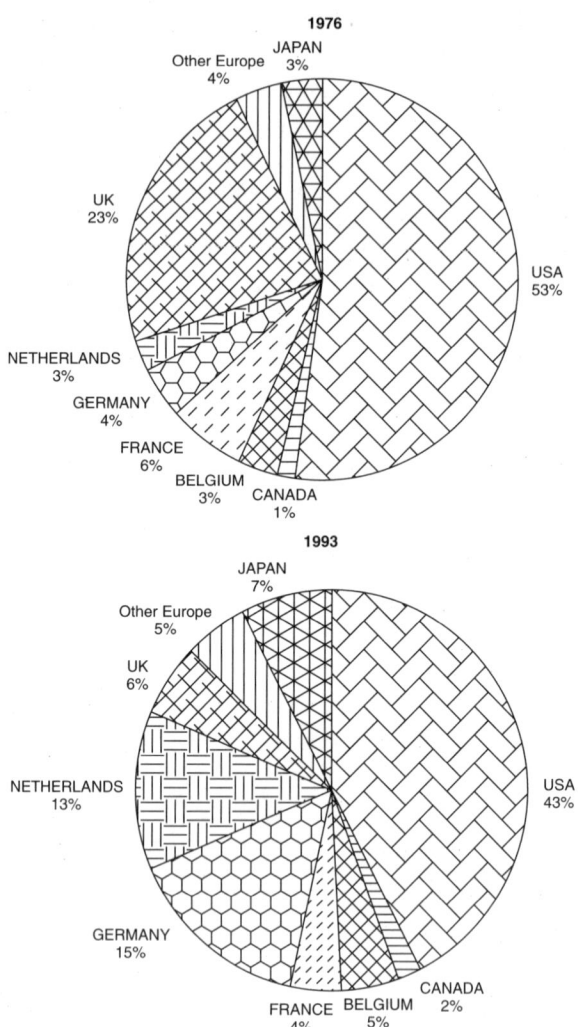

*Figure 2.1* Breakdown of receipts of technology transfer fees by reporting OECD countries, 1976 and 1993
*Source*: Kumar, based on OECD data.

cent earlier. Also, it is a bit more evenly distributed now than in the 1970s when two dominant sources accounted for 76 per cent of all royalties received.

### Emerging sources of technology

The above discussion would tend to suggest a slight trend of loosening of concentration of technology generation activity in terms of countries, with the emergence of Japan and Germany as leading sources of innovation and technology transfer. Are more countries emerging as significant sources of technology outside the OECD members?

The US patent ownership data as summarized in Table 2.3 suggests that a few countries are beginning to obtain US patents in increasing numbers over the years. These countries which include Taiwan, South Korea, Israel, Hong Kong, South Africa, Mexico, Brazil, Mainland China, Argentina, Singapore, Venezuela, India and a few East and Central European countries, collectively account for 3 per cent of US patents granted during 1990–6 compared to 1.5 per cent during 1977–82, with much of the increase coming about since 1990. However, the bulk of this 3 per cent share is accounted for by Taiwan (1.4 per cent) and South Korea (0.77 per cent), although a few countries

*Table 2.3*  Emerging sources of technology in terms of ownership of US patents, 1977–96

| Country | Patents granted during the period, and percentage share | | | | | |
|---|---|---|---|---|---|---|
| | 1977–82 | % | 1983–9 | % | 1990–6 | % |
| Taiwan | 382 | 0.10 | 2,292 | 0.24 | 11,040 | 1.43 |
| South Korea | 70 | 0.02 | 580 | 0.06 | 5,970 | 0.77 |
| Israel | 641 | 0.16 | 1,507 | 0.16 | 2,685 | 0.35 |
| Hong Kong | 272 | 0.07 | 633 | 0.07 | 1,416 | 0.18 |
| S. Africa | 491 | 0.12 | 699 | 0.07 | 787 | 0.10 |
| Mexico | 245 | 0.06 | 289 | 0.03 | 314 | 0.04 |
| Brazil | 144 | 0.04 | 212 | 0.02 | 413 | 0.05 |
| China P. Rep. | 7 | 0.00 | 142 | 0.01 | 353 | 0.05 |
| Argentina | 130 | 0.03 | 135 | 0.01 | 187 | 0.02 |
| Singapore | 17 | 0.00 | 65 | 0.01 | 337 | 0.04 |
| Venezuela | 51 | 0.01 | 122 | 0.01 | 192 | 0.02 |
| India | 56 | 0.01 | 96 | 0.01 | 204 | 0.03 |
| East and Central Europe | 3,444 | 0.87 | 2,417 | 0.25 | 1,317 | 0.17 |
| Subtotal | 5,950 | 1.51 | 9,189 | 0.96 | 25,215 | 3.26 |
| Others | 731 | 0.19 | 902 | 0.09 | 1,494 | 0.19 |
| Total | 393,629 | 100.00 | 959,368 | 100.00 | 772,927 | 100.00 |

*Source*: Kumar, based on data presented in US Patents and Trademarks Office (1997), *TAF Special Report: All Patents, All Types – January 1977–December 1996*, Washington, DC.

17

e.g. Israel, China, Singapore and India are showing an increasing trend of patenting in the US. The Latin American countries such as Brazil, Mexico, Argentina and Venezuela also own significant numbers of US patents. However, their share does not show a rising trend. The East and Central European countries also have lagged behind in technology generation and their combined share has actually fallen steadily from 0.8 per cent during 1977–82 to 0.17 per cent in the 1990s.

The rise of innovative activity in Taiwan and Korea is not only reflected in increasing numbers of US patents owned but also in emerging technology exports which are as yet small in value terms but have grown rapidly over the past few years as shown in Table 2.4. The technology exports of South Korea and Taiwan are concentrated largely in the South and Southeast Asian countries. The gradual build-up of technological effort in Asian newly industrializing countries, in particular, of South Korea and Taiwan as reflected in rising expenditures on R&D and rising numbers of overseas patents owned, is expected to result in an increase in their importance as technology exporters in the future (see Lall, 1997, for a discussion of Korean and Taiwanese technological effort). The evolution of considerable FDI outflows from some of these countries is but a reflection of the growing technological capability and created assets of national enterprises as documented in Chapter 7.

Developing countries, other than the Asian NIEs, play a negligible role in global innovative activity and clearly lag behind. Their share in global R&D expenditure or as a proportion of their domestic product has declined over the past decade (Freeman and Hagedoorn, 1992; Kumar and Siddharthan, 1997). The share of developing countries in global R&D expenditures has come down from nearly 6 per cent in 1980 to nearly 4 per cent in the early 1990s (UNESCO figures). This is despite the fact that Asian newly industrializing

*Table 2.4* Technology exports from South Korea and Taiwan, 1987–93

| Year | South Korea | | Taiwan | |
|------|--------|------------------------------|--------|------------------------|
| | *Number* | *Royalty received in million US$* | *Number* | *Values in million NT$* |
| 1987 | n.a. | n.a. | n.a. | 390 |
| 1988 | 15 | 9 | 1 | 353 |
| 1989 | 27 | 10 | 5 | 344 |
| 1990 | 50 | 35 | 3 | 785 |
| 1991 | 39 | 35 | 4 | n.a. |
| 1992 | 80 | 33 | 2 | 2,117 |
| 1993 | 105 | 45 | 2 | 1,269 |
| Total | 409 | 260 | 17 | 4,868 |

*Sources*: Kumar, based on: OECD, 1996, *Reviews of National Science and Technology Policy: Republic of Korea,* Paris; and ROC, 1995, *Indicators of Science and Technology: Republic of China 1995,* Taipei.

countries like Korea and Taiwan have greatly expanded national R&D expenditures over the past years. The trend of developing countries losing share in both R&D expenditure as well as in the global distribution of technology flows has serious implications for their place in the global technological order and would lead to a widening technological gap between the rich and poor countries. This is owing to a combination of several factors. These include the need for the governments to balance budgets as a part of the structural adjustment programmes undertaken by them which has adversely affected the public funded R&D activity. R&D activity by industry has been discouraged by the lack of protection from technology imports, heightened entry barriers and declining opportunities of adaptive R&D because of tightening patent protection. (See Kumar and Siddharthan, 1997 for a more detailed discussion.)

The growing internationalization of R&D activity of MNEs has attracted attention in recent years. The R&D activities of MNE affiliates now account for a considerable proportion of national R&D expenditure in a number of host countries, for instance, over 15 per cent in Australia, Belgium, Canada, the UK, the USA, Germany, South Korea and Singapore in the 1980s (Dunning, 1994b). However, the prospects of overseas R&D activity making a significant difference to the developing world's share of R&D activity are small. This is because the bulk of the overseas R&D activity of MNEs is concentrated in the industrialized host countries. The developing countries together account for only about 5 per cent of all overseas R&D activity of US and Japanese MNEs (Kumar, 1996b). Furthermore, even that R&D activity is concentrated in a few countries endowed with relatively superior technological infrastructure and resources among the developing countries. An empirical analysis of the factors determining the location of overseas R&D activity of MNEs found the importance of availability of technological resources and infrastructure and of host country market size. The R&D activity conducted in developing countries seemed to be more of an adaptive nature while the more creative and innovative activity was confined to industrialized countries. Therefore, it was concluded that MNEs tend to widen the gap between technologically richer and technologically poorer nations by concentrating their R&D investments in the countries already ahead in this respect (Kumar, 1996b).

The growing neglect of R&D activity by developing countries is going to have serious consequences for their ability to efficiently apply new technologies in their development. A considerable literature has argued the importance of local technological capability even for importing technology and applying it effectively in view of imperfections in the technology markets, the need for adaptations to local conditions and absorption for keeping it updated and avoiding technological obsolescence. (See, for instance, Pack and Westphal, 1986: 121; and Kumar and Siddharthan, 1997: 3, for a recent review.) The trend of developing countries losing their share in both R&D

expenditure as well as in the global distribution of technology flows would have serious implications for their growth prospects and competitiveness. Although a few Southeast Asian countries have managed to grow in the past by depending upon FDI for technological inputs for upgrading and diversification, they are increasingly becoming conscious of the constraints of this strategy and are making an effort to strengthen local technological activity (Lall, 1997).

### Trends in corporate concentration in innovative activity

Much of the innovative activity within most countries is dominated by a small number of larger corporations. Tuldar and Junne (1988) have, for instance, observed a high concentration of national R&D expenditure in a few large corporations in most of the industrialized countries. As much as 81 per cent of all Swiss national R&D expenditure in 1983 was accounted for by four companies – Ciba-Geigy, Hoffman–La Roche, Brown Boveri Corporation and Sandoz; Philips, Shell, Akzo, and Unilever accounted for 69 per cent of Dutch R&D expenditure; Siemens, Bayer, Hoechst, Daimler, and VW accounted for 22 per cent of German R&D expenditure; GM, IBM, AT&T, Ford and United Technologies accounted for 12 per cent of all the US R&D expenditure; and so on (table 6.5).

Corporate concentration in technology generation can be examined from the trends in ownership of US patents by the largest corporations. Table 2.5 summarizes the distribution of ownership of US patents owned by organiza-

*Table 2.5* Trends in the ownership of US patents held by organizations/corporations, 1977–96

| Category | Number of patents granted during the period | | |
|---|---|---|---|
| | 1977–82 | 1983–9 | 1990–6 |
| Patents owned by organizations/corporations[1] | 303,096 | 453,836 | 621,815 |
| Patents owned by top 165 organizations | 117,189 | 175,981 | 232,280 |
| Patents owned by the remaining 130,431 organizations | 185,907 | 277,855 | 389,535 |
| Patents owned by top 50 corporations | 76,345 | 113,828 | 158,999 |
| Share of top 50 in patents owned by 165 organizations, % | 65.15 | 64.68 | 68.45 |
| Share of top 50 in all organizational patents, % | 25.19 | 25.08 | 25.57 |

*Source*: Kumar, based on data presented in US Patents and Trademarks Office (1997), *TAF Special Report: All Patents, All Types – January 1977–December 1996*, Washington, DC: US PTO.
*Note*: Most, but not all, organizations are corporations.

tions and corporations by the period of grant. It is evident that the top 165 US and foreign companies and other organizations (each owning more than 1,000 patents at the end of 1996) own 38 per cent of all patents granted to the organizations. The remaining 62 per cent of patents were shared by 130,431 corporations/organizations. Within the group of 165 corporations, the top 50 account for as many as 65 per cent of patents, while they account for 25 per cent of all organizational patents. As the corporations included have not been regrouped according to the changes in ownership subsequent to the grant of patent due to mergers/ acquisitions, the actual level of concentration could be even higher. The recent spate of mergers among larger corporations and takeovers of smaller enterprises by them might, therefore, have increased concentration. The share of the 50 top corporations in the patents granted – whether in the patents owned by the top 165 or all corporations – declined slightly between 1977 and 1982 and 1983–9 (from 65.15 per cent to 64.68 per cent) but increased during the 1983–9 to 1990–6 periods (from 64.68 per cent to 68.45 per cent).

The rise in concentration of patents in the top corporations in the recent period could have taken place on account of a number of mutually reinforcing trends. The first is the rapid evolution of core technologies and their commercial potential and hence the scramble for their control. This has led to corporate restructuring worldwide as companies increasingly indulged in strategic assets seeking acquisitions. For illustration, mergers between Ciba-Geigy and Sandoz, Glaxo with Wellcome, Pharmacia and Upjohn, among others have been inspired by the need to strengthen capabilities in biotechnology related research, while the takeover of Lotus by IBM, of Word Perfect by Correl Ventura, of Intuit by Microsoft, the proposed merger between BT and MCI, among others, have been inspired by the need to strengthen competencies in the ICT technologies. In addition, larger corporations in the food, chemicals and pharmaceutical sectors have acquired a host of specialist biotechnology firms as the commercial prospects of their R&D became more apparent in recent years (see Kumar and Siddharthan, 1997, ch. 3). The corporate restructuring has in part been provoked by recent trends of deep economic integration at the regional level undertaken in the EU and North America. The national champions in the member states have been prompted to restructure their operations on a pan-regional basis. It has provoked outside corporations to take over insider companies to acquire insider status and thus evade the threat of discrimination against extra-regional goods (see Kumar, 1994b).

Table 2.6 summarizes the sales, R&D expenditure, US patents and multinationality index for the top 50 corporations owning US patents. It is apparent that all of these are giant corporations with sales in the multibillion dollar or sterling range (the largest among them being General Motors with a sales turnover of US$ 151 billion); all of them have large R&D budgets, often spending more than a billion pounds a year on R&D; all of them originated

Table 2.6 Largest fifty corporations owning US patents, 1977–96

| Rank based on total patents | Company | Home country | Sales 1994 in millions US$ | Multi-nationality index 1994* | R&D spending 1994 million US$ | US patents granted during the period | | | |
|---|---|---|---|---|---|---|---|---|---|
| | | | | | | 1977–82 | 1983–9 | 1990–6 | Cumulative |
| 1 | General Electric | US | 58,783 | 16.7 | 1,150.1 | 4,807 | 5,288 | 6,111 | 16,206 |
| 2 | IBM | US | 62,638 | 56.4 | 3,307.4 | 2,703 | 4,199 | 8,123 | 15,025 |
| 3 | Hitachi | Japan | 72,534 | 27.7 | 4,746.5 | 2,425 | 5,387 | 6,688 | 14,500 |
| 4 | Canon | Japan | 18,946 | 33.5 | 1,188.7 | 1,259 | 4,443 | 8,095 | 13,797 |
| 5 | Toshiba | Japan | 47,147 | 20.0 | 3,052.7 | 1,217 | 5,032 | 7,164 | 13,413 |
| 6 | Mitsubishi | Japan | 24,064 | 31.0 | 658.2 | 473 | 3,043 | 6,676 | 10,192 |
| 7 | Philips | NL | 34,364 | 85.0 | 2,096.4 | 2,175 | 3,936 | 3,832 | 9,943 |
| 8 | Kodak | US | 13,257 | 43.0 | 840.1 | 1,524 | 2,309 | 5,896 | 9,729 |
| 9 | AT&T | US | 73,437 | 10.8 | 3,041.4 | 2,394 | 3,313 | 3,673 | 9,380 |
| 10 | Motorola | US | 21,754 | 43.6 | 1,819.0 | 1,057 | 2,316 | 5,770 | 9,143 |
| 11 | Fuji | Japan | 10,458 | n.a. | 724.4 | 1,067 | 3,284 | 4,494 | 8,845 |
| 12 | Siemens | Germany | 53,374 | 47.3 | 4,737.0 | 2,418 | 3,399 | 2,994 | 8,811 |
| 13 | Du Pont | US | 33,291 | 42.0 | 1,023.9 | 1,990 | 2,608 | 3,608 | 8,206 |
| 14 | Bayer | Germany | 27,395 | 72.5 | 2,004.5 | 2,556 | 2,741 | 2,908 | 8,205 |
| 15 | Matsushita | Japan | 10,331 | 39.8 | 459.2 | 1,164 | 2,039 | 4,843 | 8,046 |
| 16 | Westinghouse | US | 9,177 | n.a. | 178.9 | 2,444 | 2,927 | 2,064 | 7,435 |
| 17 | NEC | Japan | 35,091 | 19.3 | 2,564.8 | 539 | 1,876 | 4,945 | 7,360 |
| 18 | Sony | Japan | 36,601 | 58.5 | 2,253.3 | 889 | 1,879 | 4,221 | 6,989 |
| 19 | Xerox | US | 17,444 | 36.7 | 875.3 | 1,764 | 1,692 | 3,522 | 6,978 |
| 20 | General Motors | US | 151,533 | 25.7 | 6,880.6 | 1,650 | 2,342 | 2,581 | 6,573 |
| 21 | Dow Chemicals | US | 19,573 | 45.3 | 1,233.2 | 1,653 | 2,628 | 2,084 | 6,365 |
| 22 | Ciba-Geigy | Swiss | 16,477 | 64.6 | 1,607.4 | 1,889 | 2,001 | 2,330 | 6,220 |
| 23 | 3M | US | 14,746 | 48.9 | 1,030.8 | 1,020 | 1,795 | 3,268 | 6,083 |
| 24 | BASF | Germany | 27,554 | n.a. | 1,209.0 | 1,383 | 1,911 | 2,483 | 5,777 |
| 25 | Fujitsu | Japan | 30,768 | 51.5 | 3,233.8 | 340 | 1,503 | 3,704 | 5,547 |
| 26 | Texas Instruments | US | 10,087 | n.a. | 673.8 | 880 | 1,550 | 3,047 | 5,477 |

| | Company | Country | Sales | | | | | | |
|---|---|---|---|---|---|---|---|---|---|
| 27 | Mobil | US | 65,284 | 58.7 | 268.9 | 1,144 | 2,332 | 1,985 | 5,461 |
| 28 | Hoechst AG | Germany | 31,316 | 64.6 | 2,125.6 | 1,543 | 1,652 | 2,252 | 5,447 |
| 29 | Sharp | Japan | 14,879 | 41.6 | 1,085.0 | 487 | 1,726 | 3,152 | 5,365 |
| 30 | Robert Bosch | Germany | 20,485 | 47.2 | 1,397.5 | 1,268 | 1,792 | 2,042 | 5,102 |
| 31 | Nissan Motor | Japan | 66,911 | 32.2 | n.a. | 1,237 | 2,188 | 1,634 | 5,059 |
| 32 | RCA | US | n.a. | n.a. | n.a. | 2,469 | 2,468 | 1 | 4,938 |
| 33 | Honda | Japan | 37,864 | 41.0 | 1,850.8 | 405 | 2,145 | 1,932 | 4,492 |
| 34 | Phillips Petroleum | US | n.a. | n.a. | n.a. | 1,644 | 1,568 | 997 | 4,209 |
| 35 | Allied Signal | US | 12,229 | n.a. | 323.7 | 745 | 1,673 | 1,726 | 4,144 |
| 36 | Ricoh | Japan | 9,488 | n.a. | 535.0 | 620 | 1,327 | 2,195 | 4,142 |
| 37 | Toyota | Japan | 92,821 | 28.1 | n.a. | 1,102 | 1,927 | 1,067 | 4,096 |
| 38 | Shell | NL/UK | 94,751 | 63.6 | 668.6 | 906 | 1,505 | 1,603 | 4,014 |
| 39 | Hughes Aircraft | US | n.a. | n.a. | n.a. | 583 | 1,056 | 2,235 | 3,874 |
| 40 | Hewlett Packard | US | 24,439 | 41.4 | 1,982.3 | 321 | 766 | 2,705 | 3,792 |
| 41 | Ford Motor | US | 125,605 | 28.6 | 5,099.0 | 668 | 1,093 | 1,915 | 3,676 |
| 42 | United Technologies | US | 20,728 | 42.5 | 956.4 | 900 | 1,347 | 1,340 | 3,587 |
| 43 | Exxon Research | US | 111,392 | 63.8 | 545.7 | 1,402 | 1,394 | 774 | 3,570 |
| 44 | Olympus | Japan | 2,352 | n.a. | 273.4 | 759 | 1,300 | 1,444 | 3,503 |
| 45 | Unisys | US | 7,235 | n.a. | 472.7 | 1,499 | 1,328 | 646 | 3,473 |
| 46 | ICI | UK | 14,059 | n.a. | 281.5 | 1,141 | 952 | 1,083 | 3,176 |
| 47 | Minolta | Japan | n.a. | n.a. | n.a. | 445 | 953 | 1,664 | 3,062 |
| 48 | Rockwell | US | 10,958 | n.a. | 582.2 | 931 | 1,186 | 922 | 3,039 |
| 49 | Procter & Gamble | US | 29,627 | 50.0 | 1,035.6 | 643 | 772 | 1,472 | 2,887 |
| 50 | American Cynamid | US | 4,422 | n.a. | 615.9 | 1,188 | 811 | 883 | 2,882 |

*Sources*: Kumar, based on the following: Sales and R&D expenditure from DTI Innovation Survey reproduced in *Financial Times*, 17 June 1995; Multi-nationality Index from UNCTAD, 1996a, *World Investment Report 1996*, Table I.12; US Patents from USPTO, 1997, *TAF Special Report: All Patents, All Types – January 1977–December 1996*, Washington, Patent and Trademark Office.
*Note*: * defined to be the average ratio of foreign component of total assets, sales and employment.

in either the US, Japan, Germany, the UK, Switzerland or the Netherlands. Finally, in terms of the multinationality index which measures the importance of foreign operations in total assets, foreign affiliate sales and employment, all have a multinational orientation. Except for AT&T, General Electric and NEC, their operations abroad accounted for 25 per cent or more of their global operations other than exports from their home bases. Many of these corporations even have more than half of their operations outside their home country, e.g. IBM, Philips, Bayer, Sony, Ciba-Geigy, BASF, Mobil, Hoechst, Shell, Exxon and P&G. A strong interaction between firm size, R&D activity, patent ownership and multinationality is thus obvious from the data presented in Table 2.6.

Not only do multinational corporations thus own a substantial chunk of technologies in the world, they often have a right of prior access to technologies generated in public funded institutions, universities and specialist small firms because of research contracts. Finally, the new technology revolution has sparked off a wave of strategic alliances which enables corporations to exploit synergies in competencies of unrelated corporations, cross license technologies and patents, and joint R&D activities. Alliance formation in joint R&D has been encouraged by the Framework Programmes of the EU which subsidize corporate R&D activity in new technologies of EU enterprises. Freeman and Hagedoorn's (1992) examination of strategic alliances compiled in the MERIT-CATI database reveals that the bulk of the technological alliances have been entered into between corporations based in industrialized countries in new core technologies. In other words, the larger corporations based in major industrialized countries may be commanding a stronger control over technology than indicated by their share of patents.

To sum up this section, the technology generation activity in the world continues to be highly concentrated in a handful of countries and a small number of relatively large corporations play a disproportionate and dominating role. While the country concentration seems to decline slightly over the years, corporate concentration at the top has apparently risen in recent years as a result of corporate restructuring provoked by new core technologies and regional economic integration in Europe and North America. The rising dominance of larger corporations may have adverse implications for the access of developing countries to new technologies on reasonable terms. Furthermore, the new norms of patent protection under the TRIPs Agreement of WTO dilute the compulsory licensing provisions by treating importing as working of patents. The potential technology suppliers, therefore, have a greater freedom to select the ways of internationalization of their operations, and in particular, use exports as a main means of exploitation of innovation (Correa, 1997). Adequate anti-trust provisions for dealing with the potential abuse of monopoly power granted by the strengthened patent system do not exist in most developing countries, however.

## Trends in international technology transfers

Data on global transfers of technology are relatively scarce. Very few countries outside the OECD group report expenditures on technology purchases from abroad on an internationally comparable basis. That leaves only supplier countries' data as a potential source for examining the trends. The OECD secretariat compiles the receipts and payments of technology fees by its member states to examine the trends in technological balance of payments of the member states. However, the trends in the global distribution of the flows are not very clear as the origin of the receipts is not provided. Hence, the trends in the technology flows are usually deduced from the inflows of FDI that are relatively easy to get from IMF, UNCTAD and OECD sources. Although FDI is a major channel of technology transfer, it is by no means the only channel. In particular, for the transfer of process technologies, arm's length licensing could be an important channel as will be seen later.

An attempt has been made here to compile some data on global flows of technology based on the technology transfer payments, license fees and royalty payments by receiving nations to supplying nations for the 1975–95 period. (See Madeuf, 1984; OECD, 1995: box 12.1, for a discussion of definitions, concepts and problems involved in using this data source.) The fees and royalties paid by developing countries have been estimated to obtain an idea of their relative importance in the global distribution of technologies.

Table 2.7 summarizes the trends in global technology transfer payments and FDI inflows over the past two decades. It is evident that both of these flows have registered an explosive growth over the past years especially since the mid-1980s (see Figures 2.2 and 2.3). The annual technology transfer payments rose from $6.8 billion in 1976 to an estimated $64.4 billion in 1995. Allowing for the OECD countries not reporting the technology receipts (especially Switzerland's and emerging NIEs'), the global magnitude of technology-related fees crossing national borders could be of the order of $68 billion. Similarly, the annual global FDI inflows rose from $22 billion in 1975 to a peak of $315 billion in 1995. However, the growth rates recorded by FDI inflows have been more impressive than those by technology transfer payments over the whole period or any of the sub-periods (see Table 2.7). This suggests that a recent spurt of FDI inflows may not have been accompanied by disembodied technology transfers in the same proportion.

A number of reasons may be responsible for the trend observed that transfers of technology may not be keeping pace with inflows of FDI. A considerable proportion of FDI inflows in the recent period has been on account of mergers and acquisitions of existing enterprises (UNCTAD, 1996a). Technology transfer in these mergers and acquisitions could be restricted to transfer of some organizational and managerial skills. The mergers and acquisitions, especially those in the industrialized countries, were led by the restructuring of businesses prompted by regional economic integration in Europe and

*Table 2.7* Global technology flows and foreign direct investments, 1975–95 (in billion US $)

| Year | Technology transfer payments | | | FDI inflows | | |
|---|---|---|---|---|---|---|
| | By all countries[1] | Estimated payments by developing countries[2] | Share of developing countries % | World[3] | To developing countries[3] | Share of developing countries % |
| 1975 | n.a. | n.a. | n.a. | 21.509 | 6.286 | 29.223 |
| 1976 | 6.82 | 1.80 | 26.40 | 27.648 | 7.050 | 25.497 |
| 1977 | 7.88 | 2.34 | 29.76 | 33.788 | 7.814 | 23.125 |
| 1978 | 9.65 | 2.67 | 27.63 | 39.928 | 8.578 | 21.483 |
| 1979 | 10.36 | 2.47 | 23.88 | 46.067 | 9.342 | 20.278 |
| 1980 | 12.48 | 3.36 | 26.93 | 52.207 | 10.106 | 19.357 |
| 1981 | 12.60 | 4.19 | 33.25 | 56.817 | 15.015 | 26.426 |
| 1982 | 10.60 | 2.16 | 20.40 | 44.472 | 13.454 | 30.252 |
| 1983 | 11.23 | 2.74 | 24.42 | 44.094 | 10.265 | 23.281 |
| 1984 | 12.01 | 3.02 | 25.17 | 48.984 | 10.500 | 21.435 |
| 1985 | 12.39 | 2.85 | 22.99 | 49.312 | 11.475 | 23.270 |
| 1986 | 16.55 | 3.52 | 21.29 | 78.283 | 14.184 | 18.119 |
| 1987 | 20.72 | 3.80 | 18.35 | 132.949 | 25.021 | 18.820 |
| 1988 | 24.43 | 4.20 | 17.21 | 158.289 | 29.718 | 18.775 |
| 1989 | 27.36 | 5.43 | 19.84 | 200.612 | 29.756 | 14.833 |
| 1990 | 33.46 | 6.70 | 20.02 | 211.425 | 31.776 | 15.029 |
| 1991 | 36.47 | 6.38 | 17.49 | 158.428 | 40.889 | 25.809 |
| 1992 | 47.30 | 7.11 | 15.04 | 170.398 | 54.750 | 32.131 |
| 1993 | 48.41 | 10.12 | 20.90 | 208.388 | 73.350 | 35.199 |
| 1994 | 56.42[4] | 14.08[4] | 24.96[4] | 225.660 | 87.024 | 38.564 |
| 1995 | 64.44[4] | 18.05[4] | 28.01[4] | 314.933 | 99.670 | 31.648 |
| Average annual growth rate, 1975–95 | 13.14 | 15.33 | | 16.40 | 16.57 | |
| Average annual growth rate, 1975–85 | 7.44 | 8.93 | | 9.56 | 7.59 | |
| Average annual growth rate, 1985–95 | 18.99[5] | 17.95[5] | | 21.19 | 24.06 | |

*Source*: Kumar, on the basis of following sources and definitions:
1 Represents total technological receipts of OECD member countries (except Denmark, Portugal, Switzerland, Ireland, Iceland, Turkey, and Greece, for which data are not available for all the years); compiled from OECD's *Basic Science and Technology Statistics*, various editions in national currencies and converted into $US with exchange rates for the respective years provided in World Bank, *World Data 1995*, CD-ROM.
2 Our estimates of technological payments by non-OECD countries to OECD countries. Estimated to be equal to the difference between technological receipts and technological payments by OECD countries.
3 UNCTC and UNCTAD.
4 Our projections on the basis of actual growth of royalty receipts of US, UK, Japan, France and Germany, 1993–5.
5 Based on the actual figures for the 1985–93 period.

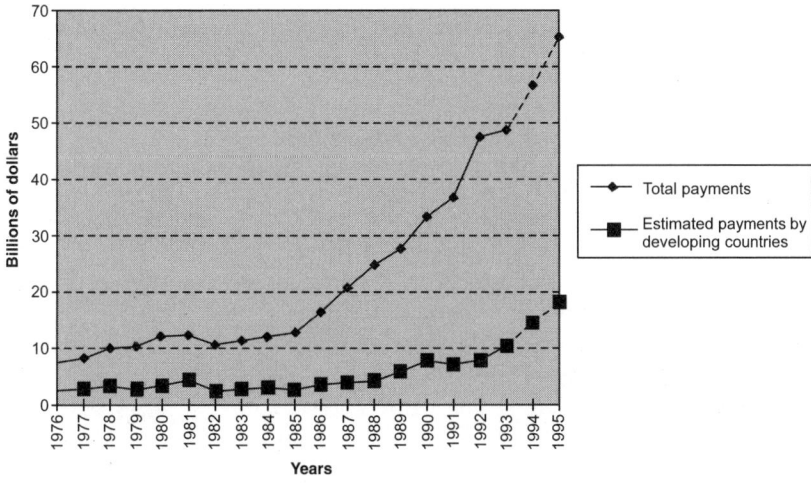

*Figure 2.2* Global technology transfer payments, 1976–95
*Source*: based on Table 2.7.

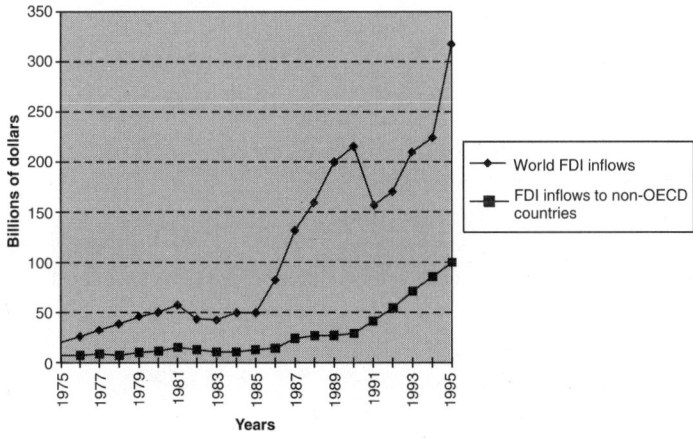

*Figure 2.3* Global FDI inflows, 1975–95
*Source*: based on Table 2.7.

North America and responses of outside enterprises to consolidate their market positions within these trade blocs, as observed earlier. A certain proportion of FDI inflows in recent years also represents debt–equity conversions in Latin America and large scale privatization of public sector enterprises in different parts of the world which also may have been accompanied by relatively little transfer of technology. Finally, the circular flow of overseas Chinese capital in Southeast and East Asia is responsible for some of the recent buoyancy in FDI inflow figures which may represent flows of largely financial capital, unaccompanied by any considerable amount of knowledge.

## The share of developing countries

Technology transfers as well as FDI inflows to developing countries have also grown over the period. The rate of growth for flows to developing countries compared to global growth is higher in the period 1985–95 (1985–93 in the case of technology flows) and one notes that technology transfers to developing countries have grown at a slower rate than overall inflows. But FDI inflows reveal a reverse pattern. This has meant that while the share of developing countries as recipients of FDI has gone up, their share in global technology transfers has come down. The share of developing countries in the two related types of flows over the years plotted in Figure 2.4 suggests that except for the 1990s the two have followed each other quite closely. In the early years developing countries' share in technology payments generally exceeded their share in FDI inflows, but an inverse tendency is apparent in more recent years. However, the recent sharp rise in the share of developing countries in global distribution of FDI inflows has to a large extent been on account of a rather heavy concentration of FDI inflows in China which has emerged as the largest host country of FDI among developing countries. The bulk of the FDI inflows attracted by China is associated with the Chinese business community in the East Asian newly industrializing countries such as Hong Kong, Macau, Singapore and Taiwan. A substantial proportion of these flows is believed to represent round-tripping of Chinese capital (rerouting itself via Hong Kong to claim the fiscal benefits meant for foreign investors) and hence may not bring new knowledge to China. Excluding the inflows to China, the share of developing countries looks comparable to those prevailing in the late 1970s and early 1980s. The long-term trend in share of technology transfers to developing countries (as well as FDI inflows, excluding the Chinese inflows) appears from Figure 2.4 to be one of decline *vis-à-vis* global transfers.

The inter-country distribution of technology transfers appears to be highly concentrated in the triad nations – the USA, the European Economic Area (EEA, comprising EU and EFTA members) and Japan. These countries are not only major sources of technology but also share the bulk of all technology inflows. Figure 2.5 depicts this phenomenon. UNCTAD studies have reported a similar importance of the triad nations in the global distribution of FDI inflows (see, for instance, UNCTAD, 1995a, figure I.6). A domination by triad countries of even greater extent in the international distribution of strategic technology alliances has been noted by Freeman and Hagedoorn (1992). As many as 95 per cent of these alliances entered into by corporations in the period 1980–9 were reportedly between enterprises from the industrialized countries, with the triad countries alone accounting for a nearly 92 per cent share. The alliances between triad and newly industrializing countries accounted for 2.3 per cent of the alliances and enterprises from all other developing countries share only 1.5 per cent.

Furthermore, it would appear that the importance of the triad countries is increasing in the global distribution of technologies originating in most of the source countries. Table 2. 8 shows that the share of triad countries in technology transferred by France and Japan increased over the past decade. Developing countries (non-OECD countries) received about 26 per cent of France's technology in 1985 but their share declined to 17 per cent a decade later. In the case of Japan, developing countries received nearly half of technology transfers in 1986; their share also declined with increasing concentration of Japanese technology transfers to the OECD countries. The data for the UK and Germany also reveal a trend similar to that observed for France – that of increasing concentration in the triad especially in the European countries. The share of developing countries has increased only within the technologies transferred by the US corporations from a rather small base at 10 per cent in 1986 to 16 per cent by 1995. Dahlman *et al.* (1995) have also noted the relative slowdown in technology transfers to developing countries. UNDP (1992: 66) similarly observed a declining share of developing countries in the capital goods trade which is the channel of international trade of embodied technology. It noted that while most developing countries are being left out of a flow that carries technological innovations and substantial scope for adaptation to local needs and conditions, technological exchanges among industrialized countries strengthen their competitiveness and make it increasingly difficult for developing countries to share the fruits of technological advance.

Even among the developing countries as a group, the distribution of technology transfers is highly concentrated in a few countries which account for a large share of technologies received from most of the source countries. Table 2.8 shows that the top five recipients accounted for more than half of all technologies received by developing countries from the US and Japan. Only

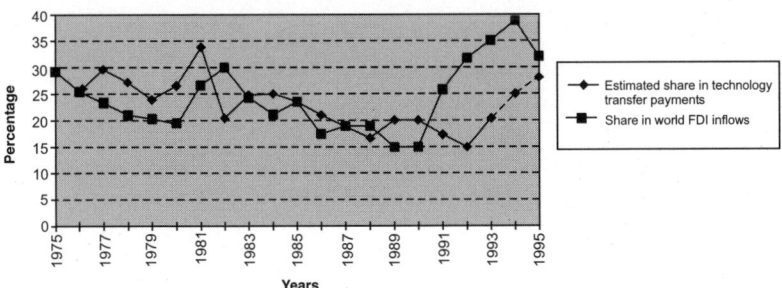

*Figure 2.4* Share of developing countries in global FDI inflows and technology transfer payments
*Source*: based on Table 2.7.

in the case of France is the distribution among developing countries more even because of technology transfers to a number of Francophone African countries. Furthermore, in the case of all the three source countries, the share of the largest five recipients has increased over the past decade. This pattern is close to that obtained in the distribution of FDI inflows with the top ten countries accounting for 81 per cent of developing countries' inflow in 1993 as compared to 66 per cent in the 1970s and 1980s. The patterns of regional focus, again like those in FDI, emerge too. Japan's technology transfers have focused on Asian NIEs. French companies have a considerable proportion of their technologies going to former French colonies in Francophone Africa. The US companies have favoured a few Latin American countries – Mexico, Brazil and Argentina, and Asian NIEs like Korea, Singapore, Taiwan, Hong Kong and Thailand.

We see, therefore, that the bulk of international technology transfers is mutually shared by the industrialized countries in the triad. The majority

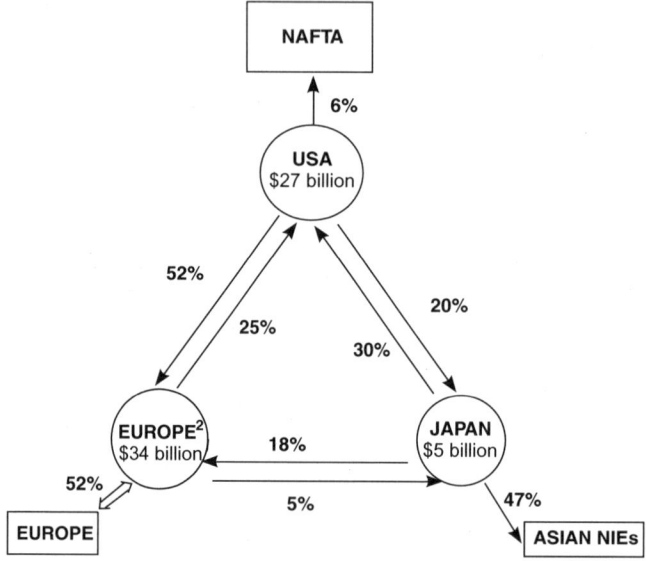

Estimated total transfers in 1995: $68 billion [1]

*Figure 2.5* Triad nations in global technology transfers, 1995
*Source*: Kumar, based on US Department of Commerce; Japan, Management and Coordination Agency; France, INPI; Deutsche Bundesbank; UK, National Statistical Office; OECD.
*Notes*: 1 Estimated value of technology transfers made by industrialized countries derived by projecting the 1993 values to 1995 at the actual growth rates achieved by the US, France, Germany, Japan and the UK in this period, and allowing for the countries for which data are not available. 2 Distribution of technology transfers made by European countries is based on actual distribution of technology receipts by Germany, France and the UK. Europe covers EU as well as EFTA members.

*Table 2.8* Inter-country distribution of technological receipts of US, France and Japan, 1985–95

| | Receipts by USA in million dollars | | Receipts by France in million francs | | Receipts by Japan in 100 million yen | |
|---|---|---|---|---|---|---|
| | 1986 | 1995 | 1985 | 1995 | 1986 | 1992 |
| All countries | 7,511 | 25,456 | 8,523 | 10,833 | 2,241 | 3,777 |
| Developed countries | 6,797 | 21,327 | 6,271 | 8,987 | 1,081 | 1,882 |
| Share in total | 90 | 84 | 74 | 83 | 48 | 50 |
| USA | | | 2,333 | 2,376 | 577 | 1,119 |
| Canada | 737 | 1,235 | 150 | 138 | 20 | 80 |
| Belgium-Luxembourg | 286 | 713 | 131 | 308 | 9 | 21 |
| France | 608 | 1,942 | | | 59 | 113 |
| Germany | 864 | 2,699 | 597 | 1,095 | 78 | 99 |
| Italy | 486 | 1,105 | 545 | 850 | 72 | 62 |
| Netherlands | 493 | 1,858 | 318 | 361 | 25 | 19 |
| Norway | 53 | 87 | 127 | 45 | 1 | 1 |
| Spain | 90 | 654 | 331 | 589 | 45 | 30 |
| Sweden | 91 | 287 | 82 | 126 | 2 | 3 |
| Switzerland | 271 | 571 | 245 | 871 | 38 | 21 |
| United Kingdom | 885 | 2,333 | 706 | 634 | 76 | 248 |
| Other Europe | 291 | 1,904 | 275 | 459 | 14 | 13 |
| Australia | 218 | 553 | 92 | 116 | 58 | 50 |
| New Zealand | 64 | 41 | 4 | 4 | 8 | 1 |
| Japan | 1,360 | 5,345 | 437 | 1,015 | | |
| Developing countries | 714 | 4,129 | 2,253 | 1,846 | 1,160 | 1,895 |
| Share in total | 10 | 16 | 26 | 17 | 52 | 50 |
| Subtotal top 5 | 368 | 2,773 | 259 | 393 | 699 | 1,263 |
| Share in developing countries | 52 | 67 | 11 | 21 | 60 | 67 |
| Argentina | 26 | 128 | 25 | 42 | 3 | 9 |
| Brazil | 25 | 311 | 17 | 18 | 38 | 22 |
| Chile | | 26 | 8 | 9 | | |
| Mexico | 109 | 414 | 42 | 52 | 13 | 24 |
| Venezuela | 15 | 93 | 65 | 25 | | |
| Other South and Western Hemisphere | 96 | 184 | 6 | 2 | 11 | 12 |
| South Africa | 71 | 112 | 24 | 74 | 17 | 27 |
| Other Africa | 9 | 31 | 652 | 469 | 171 | 170 |
| Israel | 6 | 29 | 9 | 21 | | |
| Saudi Arabia | 10 | 45 | 26 | 14 | 6 | 2 |
| Other Middle East | 14 | 12 | 266 | 101 | | |
| China | | 85 | 28 | 144 | 197 | 165 |
| Hong Kong | 33 | 289 | | | | |
| India | 54 | 48 | 65 | 57 | 44 | 50 |
| Indonesia | 8 | 56 | 55 | 23 | 152 | 131 |
| Korea, Republic of | 74 | 766 | 32 | 66 | 211 | 394 |

*Table 2.8 (Continued)*

| | | | | | | |
|---|---|---|---|---|---|---|
| Malaysia | 7 | 95 | 9 | 8 | 32 | 167 |
| Philippines | 19 | 77 | 3 | 3 | 6 | 19 |
| Singapore | 60 | 993 | 5 | 14 | 48 | 240 |
| Taiwan | 33 | 208 | 1 | 18 | 85 | 217 |
| Thailand | | 116 | 3 | 43 | 54 | 245 |
| Other Asia | 45 | 11 | 37 | 13 | 37 | 35 |

*Sources*: US Department of Commerce (Bureau of Economic Analysis), Washington, DC; France, Institute National de la Propriété Industrielle (Bureaux des Transferts Techniques Internationaux), Paris; Japan, Management and Coordination Agency (Statistics Bureau), Tokyo.

of the transactions with developing countries are actually received by a few relatively more dynamic countries and over time the concentration seems to be increasing. The recent growth in the annual flows of technology across countries, like FDI flows, appears to have benefited only a handful of countries.

## Markets versus intra-firm transactions of technology: theory, empirical evidence and recent trends

As noted in Chapter 1, the theory of international operations of firms as evolved over time with contributions from Hymer (1960), Caves (1971), Buckley and Casson (1976) and Dunning (1981), among others, considers FDI and arm's length contracts as alternative modes of foreign production. A firm exploits the revenue productivity of its intangible assets (ownership advantages), e.g. knowledge, technology, brand names, etc., abroad through FDI or within the firm (internally) if the market transactions are difficult to set up and govern. In other cases external markets are used to license the intangible assets. Thus the choice between FDI and licensing is determined by the transaction or governance costs. The higher the transaction costs, the higher the incentive to internalize the transaction (internalisation incentives) and likelihood of FDI being chosen as a mode of foreign production or technology transfer. The transaction costs are generally high for market transactions because of market failures due to the 'public-good-like' nature of technology, difficulty in making a convincing disclosure and buyer's uncertainty, problems with codification of knowledge, and risk of dissipation of brand goodwill. It must be pointed out, however, that external markets for all intangible assets are not subject to the same degree of market failure, and hence transaction costs vary. Some intangible assets such as proprietary process technologies can be profitably licensed at arm's length. Product technologies and those process technologies that cannot be codified easily or embodied in capital goods because of a high tacit component are more difficult to license. Therefore, the relative importance of licensing as a channel of technology or knowledge transfers varies across industries. Besides the characteristics of intangible assets or technology to be transferred as predicted by theory, a number of other factors may affect the choice between FDI and licensing in practice. For instance, licensing is preferred when FDI is not profitable or possible. This could be because of small size of the market or government restrictions on FDI. Licensing is encouraged when the licenser lacks experience in managing manufacturing plants abroad (Caves, 1996).

Some empirical studies have confirmed these propositions. Davidson and McFetridge (1985) in an empirical study of the choice of mode of foreign participation found certain firm- and technology-specific factors also playing a role such as age, size, past experience in technology transfer, extent of product and geographical diversification of the firm and the degree of sophistication of technological advance, among others. An OECD survey (Vickery, 1988) of licensing has shown that smaller enterprises engaged in licensing activity unaccompanied by equity ownership more than larger corporations. Local absorptive capacity and entrepreneurship are also important for licensing as the survey showed that most licensing deals arose from requests from host enterprises and that they are concentrated in industrialized countries and in the developing countries in Asia. Process technologies were found to be transferred on a licensing basis more often than product technologies. Finally, most licensing agreements included restrictive clauses such as territorial limits, exclusivity provisions and market limitations. Kumar (1987) found FDI to dominate the advertising and human skill intensive industries in an analysis of determinants of FDI and licensing across 49 Indian manufacturing industries. Licensing was important in industries where knowledge could be embodied in capital goods and those with relatively simpler technologies. Kumar (1990) found the US MNEs to prefer FDI for transfer of product technologies, and licensing for transferring process technologies.

In the period since the mid-1960s a growing proportion of international transfer of disembodied technology has taken place under arm's length contracts or licensing arrangements. This was because maturing and standardisation of a wide range of technologies led to widening of technology markets as alternative sources emerged and competition increased. A large number of host governments evolved foreign investment codes during the 1960s and 1970s and started restricting FDI inflows in an effort to reduce remittances of dividends and/or to protect domestic enterprises. Thus arm's length licensing emerged as an alternative channel of international technology transfer. This trend of rising importance of arm's length contracts as an alternative to FDI continued till the mid-1980s. It is evident from the distribution of technology receipts by major technology exporting countries reported in Table 2.9. The share of technology receipts by US corporations from their affiliates in their total technology receipts (those in which they retained a controlling stake), for instance, declined steadily from 71.36 per cent in 1975 to

Table 2.9 Composition of technology receipts, 1975–95

| Years | United States | | | United Kingdom | | | Germany | | |
|---|---|---|---|---|---|---|---|---|---|
| | Total receipts of royalties $ million | From affiliates (per cent) | From unaffiliated licensees (per cent) | Total receipts, million pounds | From affiliates (per cent) | From unaffiliated licensees (per cent) | Total receipts (for patents, innovation and processes) million DM | From affiliates (per cent) | From unaffiliated licensees (per cent) |
| 1975 | 2,643 | 1,886 (71.36) | 757 (28.64) | | | | | | |
| 1980 | 4,998 | 3,693 (73.89) | 1,305 (26.11) | | | | | | |
| 1985 | 6,121 | 4,222 (68.97) | 1,899 (31.03) | 969 | 500 (51.60) | 469 (48.40) | 1,693 | 1,559 (92.08) | 134 (7.91) |
| 1988 | 10,968 | 8,455 (77.09) | 2,513 (22.91) | 1,098 | 656 (59.74) | 442 (40.25) | 1,898 | 1,769 (93.20) | 129 (6.79) |
| 1990 | 15,507 | 12,062 (77.78) | 3,445 (22.22) | 1,420 | 1,001 (70.5) | 419 (29.5) | 2,499 | 2,336 (93.5) | 163 (6.5) |
| 1992 | 20,238 | 16,109 (79.6) | 4,129 (20.4) | 1,990 | 1,518 (76.3) | 472 (23.7) | 2,419 | 2,281[1] (94.3) | 138 (5.7) |
| 1995 | 26,953 | 21,699 (80.21) | 5,333 (19.78) | 3,339 | n.a. | n.a. | 3,116 | 2,947 (94.6) | 168 (5.4) |

Sources: US Department of Commerce, Survey of Current Business, and Monthly Report of the Deutsche Bundesbank, various issues; UN-TCMD (1993); UNCTAD (1994a).
Note: 1 belongs to 1991.

68.97 per cent in 1985. Since the mid-1980s, the share of receipts from affili-ates has constantly improved from nearly 69 per cent to over 80 per cent in 1995. The relative importance of receipts from affiliates in total receipts of technology payments varies across countries. In general, the German compa-nies tend to transfer a greater proportion of technology internally compared to American and British corporations. However, increasing importance of affiliated receipts in the more recent period is visible for the UK and Ger-many as well. Comparable figures for Japanese corporations are not available. However, the proportion of intra-firm transactions in all technology transac-tions made by Japanese corporations in number terms are available (see Chapter 6). These numbers summarized in Table 2.10 suggest that the proportion of intra-firm licensing has gone up over the 1986–92 period. In 1992 as many as 37 per cent of contracts were entered into by related firms compared to 28 per cent in 1986. One could not infer the relative tendency of Japanese corporations with regard to preference of internal mode *vis-à-vis* their counterparts from other triad nations summarized above as these figures are based on numbers rather than values.

The explanation for the decline in the importance of contractual or non-equity modes of technology transfer can be found in terms of a number of factors. These include liberalization of foreign investment policy regimes worldwide which removed the restrictions on FDI *vis-à-vis* licensing since the mid-1980s. The emergence of new core technologies – microelectronics, biotechnologies and new materials as observed earlier – has also contributed to this trend. These technologies are still evolving and are closely held. Because of their pervasive application in a wide range of sectors, they are seen by their owners as key instruments of technological competitiveness. This has prompted a wave of technological protectionism in the industrial-ized countries. Hence, companies owning them are wary of transferring them to unaffiliated parties.

The relative importance of arm's length licensing as a channel of interna-tional technology transfer varies also across industries. As observed earlier, in certain branches of industry, technology may have a high tacit component prompting the technology owners to adopt the internal or FDI mode of trans-fer. Table 2.11 shows that the proportion of technology transferred internally has varied from just 44 per cent for primary and fabricated metals to 97 per cent for non-electrical machinery in the case of technology transferred by US corporations. In the case of German companies too, primary and fabricated metals industry had least transfers internally (36 per cent). However, the tendency of internalization also varies across the source countries. The US companies, for instance, transferred 55 per cent of technology in electric and electronic equipment and 46 per cent in transportation equipment on an internal basis; the corresponding figures for German corporations were 98 and 97 per cent. As observed earlier, German companies, in general, have a much higher proportion of technologies transferred internally. The

35

Table 2.10 Share of intra-firm contracts in total technology licensing by Japanese corporations, industry by regions, 1986 and 1992 (percentages)

| Sector | 1986 | | | | | | 1992 | | | | | |
|---|---|---|---|---|---|---|---|---|---|---|---|---|
| | Asia | Latin America | North America | Europe | World | Total number of cases | Asia | Latin America | North America | Europe | World | Total number of cases |
| | (percentages) | | | | | | (percentages) | | | | | |
| Total industry | 35.9 | 26.2 | 30.5 | 13.8 | 28.1 | 2,076 | 46.3 | 25.4 | 34.3 | 20.6 | 37.0 | 3,608 |
| Manufacturing | 36.1 | 25.8 | 30.0 | 12.9 | 27.8 | 1,992 | 46.9 | 26.1 | 33.6 | 20.3 | 37.4 | 3,442 |
| Food | 25.0 | – | – | – | 25.0 | 4 | 12.0 | 0.0 | 14.3 | 7.5 | 9.0 | 233 |
| Textiles | 67.9 | 0.0 | 0.0 | 0.0 | 58.3 | 127 | 52.9 | 40.0 | 22.2 | 12.1 | 34.5 | 110 |
| Wood and pulp | 0.0 | 0.0 | 0.0 | 0.0 | 0.0 | 44 | 0.0 | – | 25.0 | 16.7 | 18.2 | 11 |
| Chemical products | 26.3 | 7.4 | 3.5 | 9.3 | 12.3 | 325 | 27.9 | 40.0 | 13.8 | 6.1 | 15.8 | 606 |
| Iron and steel | 22.7 | 22.2 | 13.3 | 0.0 | 10.8 | 83 | 10.0 | 0.0 | 33.3 | 5.1 | 13.2 | 106 |
| Non-ferrous metals | 22.9 | 50.0 | 58.3 | 8.3 | 28.1 | 64 | 42.9 | 0.0 | 35.9 | 18.8 | 34.8 | 204 |
| General machinery | 28.2 | 0.0 | 47.4 | 5.1 | 19.2 | 182 | 33.9 | 60.0 | 54.3 | 61.3 | 42.6 | 183 |
| Electric machinery | 38.8 | 55.9 | 44.2 | 24.4 | 38.7 | 511 | 71.3 | 65.0 | 26.1 | 32.4 | 58.8 | 949 |
| Transport machinery | 31.1 | 8.3 | 21.4 | 7.3 | 20.8 | 385 | 35.7 | 23.5 | 56.3 | 30.6 | 37.1 | 709 |
| Precision instruments | 50.0 | 0.0 | 61.5 | 77.8 | 53.5 | 43 | 42.9 | 100.0 | 71.4 | 63.6 | 55.9 | 34 |
| Petro and coal products | 0.0 | – | – | – | 0.0 | 1 | 0.0 | 0.0 | 37.5 | 25.0 | 19.0 | 21 |
| Other manufacturing | 28.8 | 54.5 | 39.6 | 32.1 | 33.6 | 223 | 46.5 | 30.0 | 43.1 | 54.5 | 44.6 | 276 |

Source: Urata in chapter 6 based on MITI, *Kaigai Toshi Tokei Soran* (Comprehensive Statistics of Foreign Investment), no. 3, 1991 and no. 5, 1994.

Japanese technology transfers as summarized in Table 2.10 also reveal inter-sectoral differences. As expected engineering industries which have a higher component of product technology – general machinery, electrical machinery, transport equipment and precision instruments – have a higher than average proportion of internal transfers in 1992. In the Japanese case, a relatively high proportion of transfers in textiles and non-ferrous metals industries are also conducted on an internal basis, presumably on account of their home-market--feeding nature. In the case of the food processing industry the proportion of internal transfers is very low compared to rather high ratios for the US and Germany. While the overseas expansion of Western food processing enterprises is strongly based on their brand reputations whose transfer requires high governance costs, Japanese enterprises may be transferring process technologies in this area not having built up well-known brand names as yet.

Finally, the relative importance of internal modes of technology transfer *vis-à-vis* arm's length licensing is expected to vary across the host or receiving countries depending, among other factors, upon the local absorptive capacity

*Table 2.11*  Sectoral composition of technology fees received by US and German corporations

| Industry | USA: Receipts 1989 (million dollars) | | | Germany: Receipts 1991 (million DM) | | |
|---|---|---|---|---|---|---|
| | Total | From affiliates (% share) | From unaffiliated foreigners (% share) | Total | From affiliates (% share) | From unaffiliated foreigners (% share) |
| All industries | 12,800 | 10,281 (80) | 2,519 (20) | 2,419 | 2,281 (94) | 138 (6) |
| Manufacturing | 11,212 | 9,376 (84) | 1,836 (16) | 2,284 | 2,201 (96) | 83 (4) |
| Food and kindred products[1] | 567 | 511 (90) | 55 (10) | 33 | 28 (85) | 5 (15) |
| Chemicals and allied products[2] | 2,183 | 1,707 (78) | 475 (22) | 1,232 | 1,207 (98) | 25 (2) |
| Primary and fabricated metals | 135 | 59 (44) | 76 (56) | 11 | 4 (36) | 7 (64) |
| Machinery, except electrical | 5,803 | 5,624 (97) | 179 (3) | 188 | 168 (89) | 20 (11) |
| Electric and electronic equipment | 699 | 382 (55) | 317 (45) | 508 | 497 (98) | 11 (2) |
| Transportation equipment | 250 | 114 (46) | 136 (54) | 115 | 111 (97) | 4 (3) |
| Other manufacturing | 1,576 | 978 (62) | 598 (38) | 197 | 185 (94) | 12 (6) |

*Sources*: US Department of Commerce (1992), *US Direct Investment Abroad, 1989 Benchmark Survey,* Washington, DC: GPO; Deutsche Bundesbank, *Monthly Report,* May 1992.

*Note*: 1  includes tobacco products in the case of Germany.
2  includes mineral oil refining in the case of Germany.

and technological capability in the host country, quality of local entrepreneurship, the relative level of sophistication and complexity of economic activity, the government's attitude towards FDI, and more crucially on the risk perceptions of corporations with respect to the potential host markets. Table 2.12 summarizes the patterns of transfer of technology by the US corporations in 1986 and 1995 in terms of modes. Some countries such as Japan, South Korea, Israel, India and Taiwan are known to have promoted the arm's length route of technology acquisition through selective technology imports and FDI policies in order to pursue a more autonomous path to building local technological capabilities. All of these countries have been successful in acquiring a greater proportion of technology imports on arm's length basis as is evident from Table 2.12. The proportion of external or unaffiliated imports is also high for Sweden, Chile, South Africa, Saudi Arabia, Indonesia and Thailand. Brazil, Argentina and Venezuela also imported the bulk of technology under contractual modes in 1986. Subsequently, however, they seem to have placed emphasis on internal transfers through liberalization of FDI policies; this is apparent from the sharp changes in the share of unaffiliated receipts in 1995. The Japanese data summarized in Table 2.10 enable a comparison of the proportion of internal modes across different host regions within the broadly defined industries. For instance, in the electrical machinery industry which involved 949 technology transfers in 1992 (more than any other branch of manufacturing), the proportion of internal transfers was much lower for the industrialized countries in North America and Europe than developing countries in Asia and Latin America. One possible explanation for this is that the lower local absorptive capacity in developing countries may make arm's length licensing an inappropriate mode. It is also possible that the nature of operations, e.g. in terms of market-orientation, varies across these regions.

It would appear from the above that FDI inflows seem to have been accompanied by transfers of technology in varying degrees. Some countries seem to have managed to acquire a large proportion of their technology requirements under contractual modes giving them independence of decision making crucial for charting an independent path for building local capabilities, and have rapidly emerged as technology exporters in their own right on the global scene as the East Asian countries have done. There may be others which, despite relatively large inflows of FDI, may have been unable to acquire technology and knowledge necessary to industrialize themselves and move upwards in terms of levels of competence among the league of nations. How far a nation can utilize the imported knowledge, whether as a part of the FDI package or under contractual arrangements for the augmentation of local capability to respond to the challenges of competitive industrialization, seems to depend on the quality of local entrepreneurship. Selective policies towards technology imports and other policies supportive of local technological and industrial development are most effective in the presence of a

*Table 2.12* Origin of technology fees received by US corporations, 1986 and 1995 (million dollars)

| Country | 1986 | | | | | 1995 | | | | |
|---|---|---|---|---|---|---|---|---|---|---|
| | Total | From affiliates | Share | From others | Share | Total | From affiliates | Share | From others | Share |
| All countries | 7,927 | 5,988 | (76) | 1,939 | (24) | 26,953 | 21,619 | (80) | 5,333 | (20) |
| Canada | 737 | 593 | (80) | 145 | (20) | 1,235 | 1,094 | (89) | 141 | (11) |
| Japan | 1,360 | 728 | (54) | 632 | (46) | 5,345 | 3,444 | (64) | 1,903 | (36) |
| Australia | 218 | 184 | (84) | 34 | (16) | 553 | 466 | (84) | 86 | (16) |
| New Zealand | 64 | 47 | (73) | 17 | (27) | 41 | 31 | (76) | 10 | (24) |
| Belgium-Luxembourg | 286 | 251 | (88) | 35 | (12) | 713 | 615 | (86) | 99 | (14) |
| France | 608 | 503 | (83) | 105 | (17) | 1,942 | 1,779 | (92) | 161 | (8) |
| Germany | 864 | 747 | (86) | 117 | (14) | 2,699 | 2,366 | (88) | 334 | (12) |
| Italy | 486 | 422 | (87) | 64 | (13) | 1,105 | 968 | (88) | 137 | (12) |
| Netherlands | 493 | 423 | (86) | 70 | (14) | 1,858 | 1,799 | (97) | 58 | (3) |
| Norway | 53 | 36 | (68) | 17 | (32) | 87 | 66 | (76) | 21 | (24) |
| Spain | 90 | 73 | (81) | 17 | (19) | 654 | 576 | (88) | 79 | (12) |
| Sweden | 91 | 68 | (75) | 23 | (25) | 287 | 198 | (69) | 89 | (31) |
| Switzerland | 271 | 236 | (87) | 35 | (13) | 571 | 507 | (89) | 64 | (11) |
| United Kingdom | 885 | 773 | (87) | 113 | (13) | 2,333 | 2,094 | (90) | 239 | (10) |
| Argentina | 26 | 13 | (50) | 13 | (50) | 128 | 91 | (71) | 37 | (29) |
| Brazil | 25 | 5 | (20) | 19 | (76) | 311 | 267 | (86) | 45 | (14) |
| Chile | | | | | | 26 | 10 | (38) | 16 | (62) |
| Mexico | 109 | 81 | (74) | 28 | (26) | 414 | 340 | (82) | 74 | (18) |
| Venezuela | 15 | 4 | (27) | 11 | (73) | 93 | 75 | (81) | 18 | (19) |
| South Africa | 71 | 56 | (79) | 15 | (21) | 112 | 77 | (69) | 35 | (31) |
| Israel | 6 | 3 | (50) | 3 | (50) | 29 | 12 | (41) | 18 | (62) |
| Saudi Arabia | 10 | 3 | (30) | 7 | (70) | 45 | 6 | (13) | 39 | (87) |
| Hong Kong | 33 | 30 | (91) | 3 | (9) | 289 | 232 | (80) | 57 | (20) |
| India | 54 | 5 | (9) | 49 | (91) | 48 | 9 | (19) | 39 | (81) |
| Indonesia | 8 | 3 | (38) | 5 | (63) | 56 | 22 | (39) | 34 | (61) |
| Korea, Republic of | 74 | (S) | | (S) | | 766 | 162 | (21) | 605 | (79) |
| Malaysia | 7 | 5 | (71) | 2 | (29) | 95 | 63 | (66) | 32 | (34) |
| Singapore | 60 | 40 | (67) | 20 | (33) | 993 | (S) | | (S) | |
| Taiwan | 33 | 16 | (48) | 17 | (52) | 208 | 125 | (60) | 83 | (40) |
| Thailand | | | | | | 116 | 71 | (61) | 45 | (39) |

*Source*: US Department of Commerce, *Survey of Current Business,* respective issues.

*Note*: (S) Suppressed to avoid disclosure of data of individual companies.

receptive local entrepreneurship as demonstrated by the experiences of Japan, Korea and Taiwan. Too selective or restrictive policies towards technology imports in the absence of responsive local entrepreneurship may lead to growing technological obsolescence as witnessed in the case of India during the 1970s and 1980s. For a discussion of conceptual and empirical issues concerning the interaction between technology imports and local technological development see, among others, Kumar and Siddharthan, 1997.

## Concluding remarks

The above examination of trends in technology generation and transfer in the world economy over the past two decades has shown that technology generation activity continues to be highly concentrated in a handful of advanced industrialized countries although the distribution has become slightly more even between them over time. Barring a couple of newly industrializing countries such as Taiwan and South Korea, no developing country seems to be emerging as a serious contender in technology generation. In fact the technological effort in the developing countries as a group has declined and the technology gap with the industrialized countries has widened over time rather than narrowed. The neglect of building local technological capability by developing countries will seriously affect their ability to employ even imported technologies effectively and efficiently. Therefore, policy makers in developing countries need to urgently restore the importance that technology deserves in their developmental policy and programmes.

The hold of larger corporations on the generation of new technologies has intensified with the growing scale of R&D activity, the recent trend of corporate consolidation and restructuring, different forms of inter-firm cooperation arrangements and strategic alliances, and growing industry–university links especially in the new core technologies. Furthermore, the evolution of new core technologies and the growing awareness of their commercial potential, and a general appreciation of technology in determining international competitiveness in the current age of globalization has led the governments in industrialized countries to adopt policies supporting, subsidising and protecting their national champions as a part of strategic trade policies. The increased domination of technologies by fewer corporations, coupled with the tendency for techno-protectionism, may have implications for the access and terms of transfer of technologies to developing countries especially in view of the recently strengthened and harmonized international regime of intellectual property protection. The new norms of patent protection treat importing as working of patents and provide to the patent holders greater freedom to opt for exporting to particular markets and choosing not to transfer technology. While practically all industrialized countries are equipped with effective anti-trust regulations to deal with possible abuse of the monopoly power of patent holders, most developing countries lack such effective competition policy instruments. International initiatives to establish codes of conduct for the transfer of technology and on activities of transnational corporations have failed. Adoption of effective competition policy instruments either at national levels or collectively at regional levels (*à la* European Union) to minimize the adverse impact of the possible resort to restrictive business practices by patent holders (and other anti-competitive practices especially in the context of liberalization of national economies and hence greater exposure to outside businesses) is also a possible item on the policy agenda of developing countries.

The annual magnitude of international transfers of disembodied technology in terms of royalty and technical payments has grown rapidly especially since the mid-1980s. However, their growth has not kept pace with that of FDI inflows, suggesting that recent expansion of FDI inflows has not been accompanied by technology flows in the same proportion. The inter-country distribution of global technology flows reveals that the triad nations not only generate most technology but also share among themselves the bulk of technologies transferred across countries. The technology transfers from Europe and Japan reveal also a strong regional focus. Although the share of developing countries in global technology transfers has risen in recent years, the long-term trend has been one of decline. Finally, the bulk of technology transfers to developing countries are actually received by a handful of relatively more developed and technologically more dynamic countries and this concentration is increasing. From this, the local absorptive capacity would appear to be one of the determinants of a country's ability to receive technologies from abroad. The local absorptive capacity is in turn accumulated with technological effort sustained over the years, thus reemphasizing our earlier observation regarding the importance of local technological capability for utilizing even the imported technologies.

In terms of the mode of technology transfer, the trends suggest a reversal of the growing popularity of arm's length licensing in the period from the mid-1970s to the mid-1980s. Since the mid-1980s, market or arm's length transactions have steadily lost their importance and intra-firm transfers have regained their prominence as mode of technology transfers. Although FDI or intra-firm transfers emerge as the most important mode of technology transfer, arm's length licensing is still quite a popular mode of technology transfer in a number of sectors, especially those involving process technologies. It also appears that countries with lower absorptive capacity may find it more difficult to obtain technologies on an arm's length basis. Finally, there is a considerable variation across countries in the relative importance of internal modes of technology acquired. Some countries, especially East Asian newly industrializing economies, have consciously sought technology unaccompanied by ownership as far as possible, by selective policy. As documented very well in the literature (see for instance, Kim, 1997; Kumar and Siddharthan, 1997), such policy has allowed the technology importing domestic firms independence of decision making so important for pursuing an autonomous path of expansion. The emergence of Japanese and, of late, Korean and Taiwanese corporations as multinational enterprises in their own right would not have been possible with technology acquisition under the FDI route. It must be emphasized, however, that market transfers alone do not guarantee such outcomes. They only provide necessary conditions and are effective only in the presence of responsive (and perhaps aggressive) local entrepreneurship willing to complement imported knowledge with extensive in-house technological effort on absorption, adaptation, continuous updating and eventually on innovation. In the absence of such effort, technology imports through markets may lead, as the experience of India has shown, to technological obsolescence.

# 3

# THE CHANGING GEOGRAPHY OF FOREIGN DIRECT INVESTMENT

## Explanations and implications

*John H. Dunning*

## Introduction

Like the geography of trade, the geography of foreign direct investment (FDI) both affects and is affected by the geography of production. Since the industrial revolution, each of these geographies has been in a state of flux as a result of the opening up of new territories, the discovery of new agricultural and mining products and raw materials, population growth, technological advances, and changes in the way in which economic activity is organized. However, it is doubtful whether, in the last 200 years, the geography of production has been so profoundly reconfigured over such a short period as it has between the mid-1970s and mid-1990s. Some of the more important of these changes are set out later in this chapter.

However, no less significant than the changes in the spatial distribution of economic activity have been the changing modalities by which this has been (and still is being) accomplished. Foremost among these have been the growing importance of FDI as a cross border allocator and upgrader of resources and capabilities. According to IMF and UNCTAD data,[1] between 1945 and the mid-1970s, the growth of world trade generally outpaced that of FDI, although both grew considerably faster than that of the world's gross domestic product (gdp).[2] Since the mid-1970s, however, and particularly in the last decade, FDI flows have grown at more than twice the rate of exports of goods and non-factor services;[3] and by the early 1990s, the sales of foreign affiliates of multinational enterprises (MNEs) considerably exceeded those of world wide exports[4] (UNCTAD–DTCI, 1996).

Another indicator of the growing role of FDI in global economic activity is its share of the total direct investment of the countries to which it is directed. Between the mid-1970s and 1994, this share – measured as the ratio of FDI inflows to gross domestic fixed capital formation[5] – rose from 2.2 to 3.8 per cent (UNCTAD-DTCI, 1996). Although, for countries taken as a whole, this is quite a small percentage, in at least ten developed and twenty-six developing countries it exceeded 7.5 per cent in the 1990–4 period (UNCTAD-DTCI 1996). Moreover, in both the 1970s and 1990s, the inflows were directed to the more dynamic and knowledge intensive sectors of the recipient countries (Dunning, 1985, 1993).

In this chapter, we seek to identify, and suggest reasons for, the changing geography of foreign direct investment (hereafter abbreviated to FI) between 1975 and 1980, and 1990 and 1994. In particular, we wish to pinpoint those changes which may be specifically attributed to the foreign ownership of FI, as opposed to those common to both foreign and domestic investment in the countries in which both are undertaken. In pursuing this exercise, we shall consider the main changes which have occurred in the global economy over the past two decades, and in particular, how these have affected (1) the competitive or ownership specific (O) advantages of firms, (2) the competitive or location specific (L) advantages of countries and (3) the modality by which firms coordinate their mobile O specific advantages with the immobile L specific advantages of countries (e.g. whether firms choose to buy or sell competitive advantages through intermediate product markets and/or network relationships, or whether they prefer to internalize (I) the market for these advantages).[6]

More particularly, we shall argue that it is the thrust of recent international political and economic events, and technological advances, which, through their impact on the configuration of OLI advantages facing MNEs, has been the predominant determinant of the changing geography of FI.[7] We shall also suggest that these same events have caused a fundamental – and possibly an irreversible – shift in the relative significance of the individual O, L and I variables (e.g. as set out in Dunning 1995a), as they influence both the kinds of FI undertaken by firms, and their locational pattern. In so doing, we will give especial attention, first, to the distinction between FI designed to exploit, i.e. capture the economic rent, on existing O specific advantages, and that aimed at protecting or augmenting these advantages;[8] and second, to the growing importance of high value FI, as witnessed by the significant expansion of research and development (R&D) activities undertaken by the foreign affiliates of MNEs, particularly, but not exclusively, in advanced industrial countries.[9]

Three further introductory points need to be made. First, scholarly research makes it abundantly clear that, over whatever time period one chooses to examine the geographical pattern of FI, that composition will be strongly influenced by changes in the value of three kinds of contextual variables. These are (1) the geography of the home countries of the investing firms, (2) the sectoral composition of the FI, and (3) the unique characteristics of the firms engaging in FI – including the strategies they pursue towards their foreign operations and markets. The relevance of these contextual variables will be discussed in the fourth section of this chapter.

The second point is that, for most of the chapter, we shall focus our attention on FI inflows, rather than FI outflows. However, insofar as the latter may affect the amount and geography of the total investment (TI) of a country and, hence the contribution of FI to TI,[10] it is appropriate we give some attention to FI outflows.

The third point is an important analytical one. The contribution of FI inflows to the TI of a country, and hence the extent to which FI and TI can be explained by similar phenomena, will, to a large extent, depend upon whether these inflows add to or replace investment by domestic firms (i.e. DI). Similarly, outbound FI might be at the expense of DI or complementary to it.[11] While the main thrust of this chapter is not directed to assessing the impact of FI flows on TI,[12] it will, in its discussion of the geography of FI, bear in mind the *anti-monde* or *counter-factual* situations in as much as these might affect the outcome of our analysis.

The chapter proceeds in the following way. The following section describes the changes in the geography of FI inflows between two periods, 1975–80 to 1990–4; and the third section adjusts the data for 1990–4 after allowing for changes in TI. The fourth section then offers some explanation for the changes identified. The fifth section goes on to discuss how the geography of FI will vary according to the countries making the investment; and how far changes in that geography may be explained by changes in the OLI advantages of firms from those countries, or those which are industry or firm specific. The sixth section considers the role of outward FI in affecting the geography of TI in the investing countries, and the final section summarizes the main findings of the chapter and makes a few policy remarks.

## The changing geography of foreign investment

Table 3.1 sets out the changing distribution of the flows of FDI (FI) by region and country of destination between the second half of the 1970s and the first five years of the 1990s.[13] For the first period, we have taken a six year annual average, and, for the second, a five year annual average of flows, which we believe to be a sufficiently long period to even out any 'lumpy' merger and acquisition (M&A) activity. We have chosen to use FDI flow data rather than stock figures, partly because we wish to compare the geography of foreign with that of total domestic investment (TI),[14] and it is difficult to obtain data on the stock of domestic capital for many countries;[15] and partly because such data are regularly compiled by the IMF and UNCTAD[16] and are reasonably comparable over time.

Table 3.1 suggests there have been quite significant shifts in the geography of FI over the past two decades. A comparison of the data in columns 2 and 4, as well as that of the growth of FI set out in column 5, shows that, of the

*Table 3.1* The distribution of foreign direct investment inflows by host region and country, 1975–80 and 1990–4 (millions of dollars)

| Economy | 1975–80 Annual average | % | 1990–4 Annual average | % | Index of FDI growth 1975–80 = 100 |
|---|---|---|---|---|---|
| Total inflows | 32,183 | 100.0 | 194,866 | 100.0 | 605.5 |
| Developed economies | 24,642 | 76.6 | 133,362 | 68.4 | 541.2 |
| Western Europe | 13,874 | 43.1 | 84,666 | 43.4 | 610.2 |
| of which European Union[1] | 13,190 | 41.0 | 75,688 | 38.8 | 573.8 |
| North America | 8,757 | 27.2 | 40,830 | 21.0 | 466.3 |
| of which: US | 7,895 | 24.5 | 35,615 | 18.3 | 451.1 |
| Japan | 152 | 0.5 | 1,365 | 0.7 | 898.0 |
| Other developed economies | 1,859 | 5.7 | 6,501 | 3.3 | 349.7 |
| of which: Australia | 1,271 | 3.9 | 4,352 | 2.2 | 342.4 |
| Developing economies | 7,539 | 23.4 | 57,624 | 29.6 | 763.3 |
| Africa | 810 | 2.5 | 2,905 | 1.5 | 358.6 |
| Latin America and Caribbean | 4,014 | 12.5 | 16,414 | 8.4 | 408.9 |
| of which: South America | 2,377 | 7.4 | 8,254 | 4.2 | 347.2 |
| Asia | 2,489 | 7.7 | 37,825 | 19.4 | 1,519.7 |
| of which: South, East and South East Asia | 1,971 | 6.1 | 35,361 | 18.1 | 1,794.1 |
| West and Central Asia[2] | 518 | 1.6 | 2,464 | 1.3 | 475.7 |
| Other developing economies | 226 | 0.7 | 480.0 | 0.3 | 212.4 |
| Central and Eastern Europe | 3 | neg. | 3,880 | 2.0 | ...[4] |

*Source*: UNCTC (1988) UNCTAD (1995a).
*Notes*:
1 Includes the 12 member countries of the European Union in 1994.
2 Including the Middle Eastern countries.
3 The Pacific and developing Europe.
4 Very substantial!

major regions of the world, South, East and South East Asia, Japan and Central and Eastern Europe have spectacularly increased their share of inbound investment, while the European Union[17] has posted more modest gains.

On the other hand, while the Americas, West Asia and Africa have also recorded absolute increases in inward investment – and in the case of the US, substantial increases[18] – they have lost some of their earlier attractions, relative to those of other regions or countries.[19]

Table 3.2 gives details of the leading recipients of FI in the two periods in developed and developing countries. Overall, the ten largest recipient countries accounted for 74.1 per cent of all FI flows; with just over one-half being directed to the United States, the United Kingdom, France and the

*Table 3.2* The largest recipients of inward FDI, 1975–80 and 1990–4 (annual averages)

| | Developed countries | | | | | | Developing countries[4] | | | | | |
|---|---|---|---|---|---|---|---|---|---|---|---|---|
| | 1975–80 | | | 1990–4 | | | 1975–80 | | | 1990–4 | | |
| Country | Country | $m | %[1] | Country | $m[1] | % | Country | $m | %[2] | Country | $m | %[2] |
| | United States | 7,894.0 | 32.0 | United States | 35,615.0 | 26.7 | Brazil | 1,835.8 | 24.4 | China | 16,064.8 | 27.9 |
| | UK | 5,795.4 | 21.1 | UK | 17,756.2 | 13.3 | Mexico | 1,023.5 | 13.5 | Singapore | 6,384.4 | 11.1 |
| | France | 2,127.3 | 8.6 | France | 17,571.2 | 13.2 | Malaysia | 524.3 | 7.0 | Mexico | 4,332.0 | 7.5 |
| | Netherlands | 1,276.6 | 5.2 | Spain | 9,480.0 | 7.1 | Singapore | 502.0 | 6.7 | Malaysia | 4,243.8 | 7.4 |
| | Australia | 1,271.4 | 5.2 | Belgium | 9,079.0 | 6.8 | Egypt | 376.1 | 5.0 | Argentina | 3,191.8 | 5.5 |
| | Belgium[3] | 1,203.1 | 4.9 | Netherlands | 7,027.6 | 5.3 | Iran | 315.5 | 4.2 | Thailand | 2,197.8 | 3.8 |
| | Germany | 1,052.6 | 4.3 | Germany | 5,476.0 | 4.8 | Indonesia | 289.9 | 3.8 | Indonesia | 1,871.2 | 3.2 |
| | Spain | 970.5 | 3.9 | Canada | 5,215.0 | 3.9 | Hong Kong | 241.1 | 3.2 | Hong Kong | 1,596.8 | 2.8 |
| Top 8 | Top 8 | 20,991.6 | 85.2 | Top 8 | 10,7220.0 | 80.4 | Top 8 | 5,108.2 | 67.8 | Top 8 | 39,882.6 | 69.2 |
| All | All | 24,642.0 | 100.0 | All | 133,361.6 | 100.0 | All | 7,539.1 | 100.0 | All | 57,623.8 | 100.0 |

*Notes:* NB: In the years 1975 to 1980, the ten largest recipients of FI identified in the above table accounted for 74.1 per cent of all FI inflows, and in 1990–4 they accounted for 68.8 per cent.
In the former period, Japan accounted for only 0.6 per cent of all inflows into developing countries; and in the latter for 1.0 per cent.
1 Of all developed country investment.
2 Of all developing country investment.
3 And Luxembourg.
4 Bermuda was, in fact, ranked higher – 6th, but we have excluded the tax haven countries from our rankings.

Netherlands. By the first half of the 1990s, this geographical concentration index had decreased to 66.1 per cent, although four leading countries – the United States, the United Kingdom, France and China – still accounted for 45 per cent of the total FI flows.

Table 3.2 also shows that, among the most significant changes in the distribution of inbound FI between countries over the past two decades have been, first, the rise of China to the fourth largest recipient; and second, the declining drawing power of Brazil and one of the leading oil exporting countries in the 1970s – Egypt. There has been some shifting in the ranking of the leading recipients of FI. Among the developed countries, for example, Spain and Canada have both increased their share over the past two decades, while, of the developing countries, Singapore, Argentina and Thailand have strengthened their positions.

It is worth noting that there are some leading developed and developing economies which do not receive as much FI as one might have expected. Japan is the most obvious example; it accounted for only 0.6 per cent inflows into the developed countries in 1975–80 and 1.0 per cent in 1990–4. Of the larger European countries,[20] Italy received only one-quarter the share of France in both periods, while some of the more populated newly industrial countries (NICs) of Asia – Korea, Taiwan and the Philippines – attracted only modest (though increasing) amounts of FI. A more detailed description of the changes in the geographical distribution of FI is set out in Appendix Table 3.A1.

The case of India – and even more spectacularly, that of some Central and East European countries – is interesting in that, although their share of total inward FI remains very small, the rate of growth of investment directed to these countries over the last two decades has been well above average. As Table 3.A1 shows, the same is also true of some smaller developed and developing countries.[21] On balance, however, there has been a slight trend towards a more even distribution of FI. In the case of developed countries, the standard deviation around the mean amount of inbound FI of $1,265 million in the period 1975–80 was $1,995 million; and for developing countries around a mean of $195 million, it was $354 million. The corresponding means and standard deviation for the average annual amount of FI in the 1990–4 period were $6,609 million and $8,848 million for developed countries, and $950 million and $1,335 for developing countries.

## How far do the changes in the geography of FI reflect those in the geography of total investment?

In seeking to explain changes in the geography of FI which reflect its distinctive ownership, it is first necessary to see how far these correspond to changes in the geography of investment by domestically owned firms, i.e. DI. For, if the changes in each were exactly the same, then it could be argued that the 'foreign-ness' of investment had no impact on its geography – and that this geography could be explained by the factors influencing DI.

The nearest domestic counterpart to FI we have, which is both regularly published for most countries and is reasonably comparable, is that of gross fixed capital formation (GFCF). This, in fact, includes the capital expenditures of foreign owned firms, and so we shall refer to it as total investment (TI).[22] In Table 3.3 we attempt to do two things. First, we compare the geography of FI and TI in 1975–80 and 1990–4. Second, we adjust the FI data for 1990–4 to take account of the changes in the distribution in TI; and hypothesize that the difference between these changes and the actual changes in FI represents that part of the latter which can be specifically attributable to its non-resident ownership.[23] For reasons we shall outline shortly, this is a heroic assumption, not least because FI in one period might affect TI in another period. But, this procedure does take us some way in isolating one of the components of the changing geography of FI.

The data in Table 3.3 reveal two main things. First, the geography of FI and TI, although broadly similar, does exhibit noticeable differences. In 1975–80, for example, as shown by the FI/TI ratios in column 3 and by the actual percentage of TI accounted for by FI, there was an above average FI intensity,[24] in Western Europe, in other developed countries, in Latin America and in Africa. By contrast, the share of worldwide TI accounted for by Asia exceeded that of its share of FI – and, in Japan's case, by a very large margin. For the US and Canada, the share of both FI and TI was about the same.[25]

The situation for 1990–4 continued to show that the FI intensity ratio was decidedly higher in the case of developing economies than in that of developed economies, but the absolute difference narrowed. Partly, this reflected the much greater share of Japan in the world's GFCF – while that country continued to attract about the same (and very small) proportion of FI flows. Partly, too, it mirrored the substantial increase in the share of FI in the TI of South and East Asian countries; and, most spectacularly of all, in China, where the ratio of FI inflow to GFCF rose from a miniscule amount in 1975–80 to 10.8 per cent in 1990–4.[26]

The second conclusion to be drawn from Table 3.3 is that the rates of growth of FI and TI between the second half of the 1970s and the first half of the 1990s differ between countries. Column 9 identifies those regions and/or leading recipient countries in which FI has risen faster than TI – e.g. Africa and parts of Asia – and those in which the reverse is the case, e.g. the Americas. An alternative way of looking at the same phenomenon is to track directly changes in the FI/GFCF ratios of regions and countries. These are set out in Table 3.4. Here, we have classified the countries[27] by their main regions and related changes in the FI/GFCF ratios to the average for the region of which they are part. The data in the table are largely self-explanatory. The most substantial increases in the FI intensity ratio in

Table 3.3 The changing geography of foreign and total investment,[1] 1975–80 and 1990–4

| Economy | 1975–80 FI % | 1975–80 TI % | 1975–80 FI/TI | 1990–4 FI % | 1990–4 TI % | 1990–4 FI/TI | 1990–4/1975–80 FI Ratio | 1990–4/1975–80 TI Ratio | 1990–4/1975–80 FI/TI | 1990–4 FI(A) % | 1990–4 TI % | 1990–4 FI(A)/TI |
|---|---|---|---|---|---|---|---|---|---|---|---|---|
| Developed economies | 78.3 | 84.4 | 0.93 | 69.6 | 73.9 | 0.94 | 5.36 | 1.81 | 2.96 | 74.6 | 73.9 | 1.01[2] |
| Western Europe | 44.1 | 35.3 | 1.25 | 43.4 | 43.3 | 1.03 | 5.94 | 2.48 | 2.40 | 45.0 | 43.3 | 1.04 |
| North America | 27.8 | 28.1 | 0.99 | 21.6 | 29.7 | 0.73 | 4.67 | 2.18 | 2.14 | 25.3 | 29.7 | 0.85 |
| of which: US | 25.1 | 24.9 | 1.01 | 18.8 | 26.3 | 0.71 | 4.52 | 2.18 | 2.07 | 22.1 | 26.3 | 0.84 |
| Japan | 0.5 | 19.2 | 0.03 | 0.8 | 28.7 | 0.02 | 9.87 | 3.08 | 3.20 | 0.1 | 28.7 | 0.00 |
| Other developed economies | 4.8 | 1.7 | 2.73 | 3.7 | 1.9 | 1.95 | 4.73 | 2.27 | 2.08 | 4.3 | 1.9 | 2.26 |
| of which: Australia | 4.0 | 1.7 | 2.31 | 2.5 | 1.9 | 1.17 | | | | | 1.9 | 0.95 |
| Developing economies | 21.0 | 14.9 | 1.41 | 30.1 | 25.8 | 1.17 | 8.62 | 3.57 | 2.41 | 24.5 | 25.8 | 0.95 |
| Africa | 0.9 | 0.8 | 1.07 | 1.5 | 1.0 | 1.53 | 10.38 | 2.47 | 4.20 | 1.6 | 1.0 | 1.60 |
| Latin America | 13.7 | 7.3 | 1.87 | 8.9 | 5.5 | 1.63 | 3.91 | 1.54 | 2.54 | 14.8 | 5.5 | 2.76 |
| of which: South America | 8.0 | 5.2 | 1.52 | 4.6 | 3.3 | 1.41 | 3.47 | 1.28 | 2.70 | 9.1 | 3.3 | 2.80 |
| Asia | 6.5 | 6.8 | 0.95 | 18.8 | 19.4 | 0.97 | 17.38 | 5.87 | 2.96 | 8.2 | 19.4 | 0.42 |
| of which: South, East & South East Asia | 6.2 | 6.8 | 0.92 | 17.7 | 16.6 | 1.07 | 17.19 | 5.06 | 3.40 | 9.0 | 16.6 | 0.54 |
| Other developing economies | 0.7 | 0.7 | 1.0 | 0.3 | 0.3 | 1.0 | 2.59 | 0.89 | 2.92 | 0.9 | 0.5 | 3.00 |
| Central and Eastern Europe | neg. | | | 0.2 | | | ∴[3] | | | | | |
| All countries | 100.0 | 100.0 | 1.01 | 100.0 | 100.0 | 1.0 | 5.88 | 2.07 | 2.84 | 100.0 | 100.0 | 0.99 |
| Value US$billions | 32.2 | 1584.3 | | 189.1 | 3271.8 | | | | | | | |

Source: UNCTC (1988), UNCTAD (1995a) and UN (various).
Notes:
1 Gross fixed capital formation.
2 Adjusted FI (FI (A)) is the annual average FI in the 1975–80 period multiplied by the rate of growth of FI between 1975–80 and 1990–4, divided by the growth of TI over the same period.
3 Very substantial.

Table 3.4 Changes in the ratio of FI flows to gross fixed capital, 1975–80 and 1990–4

### Developed countries

**Western Europe** — Average change in ratio (2.6 to 5.0)

| | Below average | Around average | Above average |
|---|---|---|---|
| Germany | 0.80 to 0.38 | | |
| Ireland | 6.70 to 1.24 | | |
| Italy | 0.80 to 1.52 | | |
| Norway | 3.20 to 3.30 | | |
| UK | 8.40 to 10.46 | | |
| Greece | 5.40 to 7.94 | | |
| Austria | | 0.90 to 1.80 | |
| Netherlands | | 4.50 to 12.20 | |
| Belgium Lux. | | | 5.80 to 22.28 |
| Denmark | | | 0.30 to 8.68 |
| Finland | | | 0.40 to 4.18 |
| France | | | 1.90 to 7.08 |
| Portugal | | | 1.50 to 10.68 |
| Spain | | | 2.80 to 10.14 |
| Sweden | | | 0.50 to 11.06 |

**North America** — Average change in ratio (2.0 to 3.0)

| | Around average |
|---|---|
| Canada | 1.70 to 4.50 |
| US | 2.00 to 4.14 |

**Other developed countries** — Average change in ratio (0.5 to 0.8)

| | Around average |
|---|---|
| Japan | 0.10 to 0.20 |
| Israel | 1.10 to 2.68 |

### Developing countries

**Asia** — Average change in ratio (2.6 to 5.0)

| | Below average | Around average | Above average |
|---|---|---|---|
| Hong Kong | 4.20 to 6.76 | | |
| Rep of Korea | 0.40 to 0.70 | | |
| Indonesia | 2.40 to 3.54 | | |
| Thailand | 1.50 to 2.86 | | |
| Cyprus | 11.10 to 6.22 | | |
| Oman | 8.90 to 6.96 | | |
| Malaysia | 11.90 to 22.44 | | |
| Sri Lanka | | 2.60 to 4.62 | |
| Taiwan | | 1.20 to 3.02 | |
| Singapore | | 16.60 to 28.40 | |
| Papua New Guin. | | 8.70 to 13.44 | |
| India | | | 0.10 to 0.48 |
| Pakistan | | | 0.90 to 3.56 |
| China | | | 0.80 to 11.64 |
| Philippines | | | 0.90 to 6.16 |

**Latin America** — Average change in ratio (2.5 to 8.0)

| | Below average | Around average | Above average |
|---|---|---|---|
| Bolivia | 2.30 to 3.00 | | |
| Brazil | 3.90 to 3.23 | | |
| Dominican Republic | 5.00 to 8.38 | | |
| Uruguay | 15.70 to 4.30 | | |
| Mexico | 3.60 to 7.38 | | |
| Paraguay | 3.90 to 7.09 | | |
| Guatemala | 7.90 to 5.64 | | |
| Peru | | 2.00 to 7.78 | |
| Chile | | 4.20 to 9.98 | |
| Costa Rica | | 6.00 to 12.84 | |
| Trin. and Tob. | | 10.70 to 42.22 | |
| Columbia | | | 2.20 to 10.98 |
| Ecuador | | | 1.80 to 9.18 |
| Venezuela | | | −0.90 to 8.22 |
| Argentina | | | 2.10 to 29.34 |
| Jamaica | | | 1.60 to 11.66 |

**Africa** — Average change in ratio (2.5 to 8.0)

| | Below average | Around average | Above average |
|---|---|---|---|
| Côte d'Ivoire | 2.50 to −2.46 | | |
| Gabon | 3.20 to −1.28 | | |
| Kenya | 4.40 to 1.08 | | |
| Egypt | 7.10 to 6.70 | | |
| Tunisia | 6.10 to 7.04 | | |
| Swaziland | 18.80 to 33.86 | | |
| Liberia | | 18.70 to 55.44 | |
| Nigeria | | | 0.50 to 29.66 |

Sources: UNCTC (1988), UNCTAD (1995) and UN (various).

Note: Data for Central and Eastern Europe are not presented as there was virtually no FI in any of these countries in the 1975–80 period.

national economies, relative to the average for their regions[28] are recorded by the Scandinavian countries, Portugal and France in Europe; Morocco, Namibia, Nigeria and Zambia in Africa; China, India and the Philippines in Asia; and Argentina, Paraguay and the Dominican Republic in Latin America. Representative of the countries where the share of GFCF has either fallen or not kept pace with that of the average for their region are Germany, Ireland and the UK in Europe; Botswana, Egypt, Ghana, Kenya and Malawi in Africa; Hong Kong and Korea in Asia; and Brazil in Latin America.

The data set out in Tables 3.3 and 3.4 also suggests that, while the share of FI in TI has fluctuated between 1975 and 1980 and between 1990 and 1994, there is a reasonably strong positive association between the absolute growth of the two variables. The Pearson coefficient of the correlation for the 55 countries identified in Table 3.4 was 0.45; excluding India, Brazil and Japan, the coefficient rises to 0.68. This being so, and bearing in mind that the purpose of this chapter is to describe and explain the distinctive geography of FI, we have adjusted our FI data for 1990–4 set out in Table 3.3 (column 4) to take account of that part of its growth which may be due to factors other than its foreign-ness. This we have done (as the formula at the bottom of Table 3.3 describes) by deflating the growth rate of FI by that of TI. In column 10, we give the adjusted values (FI(A)) of FI flows for 1990–4, and in column 11 we set out the geographical distribution of these flows. Comparing these data with those in column 3 and 4, it can be seen, first, that the changes in the distribution in FI(A) over the past twenty years are less marked than those of FI; and that, second, there are significant differences in adjusted, cf. unadjusted, FI intensity ratios (cf. columns 12 and 9).

We are, of course, aware of the limitations of such a procedure, not least that – as we shall explain in more detail – the growth of TI might, itself, be affected by FI, both directly (as FI is part of TI) and indirectly, as FI might increase (or, indeed, lower) TI. The interaction between FI and TI is likely to be most pronounced where the FI/TI ratio is highest or increasing the most, and/or where TI is concentrated in the key sectors of the host economy. Notwithstanding this caveat, however, we believe that the geographical distribution of the FI(A) set out in the final column of Table 3.3, when, compared with the average FI for 1975–80, gives a truer picture of the changes which are specifically due to the ownership of that investment; and the rest of this paper builds upon this belief!

Adjusting the FI flow data for 1990–4 also has a marginal effect on its distribution among the leading recipient countries. Table 3.2 showed that the ten leading recipient countries accounted for 67.8 per cent of total unadjusted FI flows; normalizing these data for the percentage growth of TI (see Table 3.3), this percentage rises to 72.0 per cent. Six of the eight developed countries and seven of the developing countries which attracted the most FI in

1990–4 also attracted the most FI(A) in the same period. However, of the former group, Australia and Italy replaced Germany and Canada as the seventh and eighth ranking; and of the latter, Indonesia was replaced by Taiwan.

## Explaining the changes in the geography of FI

The main reasons why changes in the spatial distribution of FI and TI might be different are, first, that their respective owners or managers may perceive investment opportunities differently, and/or act on any particular perception differently; and second, that they may react differently to changes in exogenous (i.e. extra-firm) variables which affect either the desirability of the investment or the locational opportunities for its deployment.

In the language of the eclectic paradigm of international production (Dunning 1993, 1995a), these differences essentially reflect (1) different competitive or ownership (O) specific advantages of domestic and foreign owned enterprises; (2) different strategies of the two groups of firms in organizing and deploying these advantages in the light of the competitive or locational (L) specific advantages of particular countries;[29] and/or (3) the way in which – over time – (1) and (2) impact on the L advantages of the countries. Thus, for example, foreign MNEs might increase their share of TI in a particular host country either because their O advantages, relative to those of indigenous firms, are increasing; or because of their desire to protect, or augment, their existing O advantages by acquiring the assets of these firms. Alternatively, because of the character of their O advantages, or their particular innovatory and organizational strategies, the elasticity of response of foreign firms, to a change in the attractiveness of a particular location might be higher than that of indigenous firms. Or, indeed, the incentives offered by countries seeking to attract economic activity might discriminate in favour of foreign firms.

By the same token, the rate of growth of TI might exceed that of FI if there are restrictions placed on new FI, e.g. new local content requirements; or that, stimulated by the presence of foreign subsidiaries, domestic firms upgrade their competitiveness and invest proportionately more than the subsidiaries; or where the attractions of concluding cross-border licensing agreements and other forms of collaborative agreements favour indigenous, rather than inbound, investors.

If the changing balance of the O and I specific advantages of foreign and domestically owned firms, and the differential impact of changes in the L specific advantages of countries on their behaviour, are the main explanations of the changing geography of FI (as distinct from TI), the question next arises: 'What has caused these changes over the last 20 years; and, in particular, are there any differences in the configuration between the O advantages of firms and the L advantages of countries in the 1990s as compared with those in the later 1970s?'

Let us consider first the main changes in the global economic scenario over the last twenty years. Although there is no clear watershed date which initiated the changes, we believe that the beginning of the regimes of Margaret Thatcher in the UK and Ronald Reagan in the US offers as good a dividing line between the two eras of FI as any other. As these changes are fairly well known, we will do no more than set them out in Table 3.5.

*Table 3.5* Some features of the changing world economic scenario: late 1970s to early 1990s

---

1 The renaissance of the market economy, as the predominant form of economic organization by most countries. Along with such renaissance has come the liberalization, deregulation and privatization of markets which have lowered the artificial or man-made costs of the movement of goods, services and assets

2 In the light of (1), the reorientation of the macro-economic and macro-organizational philosophies of national governments towards more market-enabling and less market-distorting policies – at least as far as wealth creating activities are concerned

3 The coming on-stream of a new generation of technological advances, and particularly those in telecommunications, which has hastened the trend towards knowledge-based capitalism; and which, for the most part, has lowered the costs of trasversing space assets goods and people

4 In the light of 1–3 above, the promotion of many regional economic schemes (notably the completion of the internal market in Europe, NAFTA and ASEAN)

5 In the light of 1–4 above, the growing competition among firms – including firms of different nationalities. In a real sense, regional and global competition for resources and markets is replacing national competition – particularly in international industries

6 In the light of 1–5 above, firms have reorganized and restructured the range and composition of their value-added activities. Sometimes this has resulted in down-sizing and disinternalization of intermediate product markets; in other cases, it has prompted more mergers and acquisitions so that firms can better capture the benefits of economies of scale and scope; and acquire competitive-enhancing assets

7 Partly as a consequence of 1–6, and the emergence of China, the newly industrializing countries (NICs) and some Central European economies, as important actors in the global market economy – each of which has different propensities both to be invested in – by foreign firms – and to engage in outward foreign investment, the geography of FI has undergone some major changes since the mid-1970s

---

We have done so sequentially, although we fully appreciate that many of these changes are interdependent among each other. It is, however, their combined effects on (1) the competitive advantages of enterprises and (2) the competitive advantages of countries, and on how these impact on the organization and geography of FI to which we wish to give special attention.

## *The changing character of the O specific advantages of firms*

Prior to the late 1970s, the main O advantages possessed by foreign investors over their indigenous competitors (some of which, themselves, were MNEs) were their privileged access to specific intangible assets, e.g. technology, management and organizational skills, markets, etc. Elsewhere, we have termed these Oa advantages (Dunning, 1993). For the most part, too, the core competencies of MNEs reflected the resource endowments of their home countries, rather than those which stemmed from the countries in which they made their foreign investments. Few MNEs, at that time, practised globally integrated production or marketing strategies, although, within the European Community, there was some product or process specialization and intra-firm trade, particularly among US affiliates. While the propensity of MNEs to increase the foreign component of their total value added activities was leading to more sequential or efficiency seeking FI (Kogut, 1983), the majority of firms pursued multi-domestic strategies towards their global operations; and most foreign subsidiaries outside the resource based and labour intensive manufacturing sectors operated on a 'stand alone' basis and traded little with each other.

In the 1970s scenario, insofar as the geography of FI reflected the distribution of O specific advantage among firms of different nationalities, these tended also to reflect the structure of the natural and created assets, and the markets, of their home countries. There is ample empirical evidence, both of the geography of outward FI, and of that of recipient countries attracting inward FI, to support this assertion.[30] Moreover, these same O advantages help to explain at least some of contemporary FI – and particularly European and US FI in Japan (Dunning, 1996a), Japanese FDI in the US (Hennart and Park, 1994), and that by third-world MNEs (Dunning, van Hoesel and Narula, 1998).

Over the last two decades, the nature and character of the O advantages of MNEs has changed. In general, their country of ownership has become a less important determinant, and the degree of their multinationality – which is a firm-specific variable – a more important one. Moreover, although a component of the core competencies of firms continues to be the exclusive or privileged possession of specific assets, increasingly, it is the way in which these assets are governed and coordinated with the assets of other firms, and with the L specific endowments of countries or regions in which they operate, which is determining the extent and pattern of O specific advantages, and the location of their creation and deployment.

In the emerging age of alliance capitalism, a firm's O specific advantages rest not only on its in-house capabilities to pursue cost-minimizing strategies,[31] and/or upgrade its core assets and innovate new ones, but its willingness to seek out and augment assets complementary to its own, by networking and other cooperative arrangements with its suppliers, competitors and customers (Dunning, 1995a). The benefits of the common governance of inter-related cross-border operations – which have been variously described as transaction cost minimizing (Ot) or sequential advantages – are particularly well demonstrated in the case of knowledge or learning based firms with multiple home bases (Porter, 1990). It is these firms which are best positioned to take advantage of the liberalization of markets, regional integration and new technological advances; and to foster an international division of labour based, not on the spatial disposition of natural resources, but on that of created assets. Often such a geography has very different implications for the location of TI and FI; as the critical determinants of the former, e.g. the quality and cost of land and unskilled, or semi-skilled, labour are replaced by the availability and costs of a supportive physical and human-based infrastructure, and the ease of access to foreign markets.

At the same time, the accelerating economic development of some NICs, particularly in South and East Asia, has widened the constituency of investing firms, and changed the character of their O specific advantages. Clearly, although there are similarities between the core competencies of Japanese and third-world MNEs – as compared with European and US MNEs – there are also many differences. Some of these will be identified in the next section of the chapter, but the point we wish to make at this juncture is that different O advantages may bring with them different locational consequences; and that part of the reason for the changing geography of FI may rest in the different O specific advantages of the new generation of outward investors.[32]

In, perhaps, no other area of value added activity have the changing O advantages of firms more affected the geography of FI than that of research and development (R&D). Prior to the 1970s, there was comparatively little R&D – and particularly fundamental R&D – undertaken by MNEs outside their home countries;[33] although there were some exceptions, e.g. MNEs from the UK, Canada and some smaller European nations, and those in sectors such as food, beverages and pharmaceuticals, where local supply or demand conditions made it desirable for some R&D to be decentralized. However, even in these cases, the predominant motive for foreign based innovatory activity was to adapt home based R&D and create peripheral products and processes – i.e. what Kuemmerle (1996: 9) refers to as 'home based exploiting R&D'.

Over the last two decades, not only has the percentage of innovatory activity undertaken by MNEs outside their home countries risen sharply,[34] but the reasons for conducting such activity – at least in industrialized countries – have broadened to include both the creation of new core products and/or processes, and the acquisition of R&D facilities necessary to advance the productivity of domestic R&D or, as Kuemmerle (1996) puts it, to 'augment home based R&D'. Such a strategy has very different locational implications from that of exploiting home-based R&D advantages; and, indeed, it suggests a more concentrated geographical pattern of FI to take advantage of clusters of technological expertise and expense. Certainly, what little empirical evidence we have[35] supports this contention; and indicates that, in spite of the increase in the share of FI directed to developing countries over the past decade or so, the share of R&D activity so directed by MNEs may well have fallen.[36] This is not to say that MNEs are undertaking less R&D in developing countries – both US and Japanese data belie this – but that innovatory related FI has not followed the same locational pattern as has that of other kinds of FI in the manufacturing and service sectors.

### The changing character of the L advantages of countries

Not only has the widening and deepening of MNE activity from market-seeking and natural resource seeking FI to embrace efficiency and strategic asset seeking FI[37] required firms to reappraise their locational strategies, but world economic events have had a direct impact on the structure and character of the competitive advantages of countries.

In the 1970s, the locational advantages of countries lay primarily in their favoured possession of natural resources and unskilled or semi-skilled labor, and their ease of access to markets for finished products. It was such L specific endowments which firms were looking to add value to their mobile O specific advantages. Traditional location theory tended to classify the variables affecting the siting of foreign production into three groups – those which were cost related, revenue related and profit related (Guisinger and associates, 1985). Government imposed taxes or subsidies might be deployed to affect one or other of these variables (UNCTAD–DTCI, 1995a), the significance of which depends on the type of FI and the country in which it was being made. Most of the variables affecting FI were common to TI, although some, notably investment incentives and some non-pecuniary requirements were specific to foreign investors (UNCTAD–DTCI, 1995a).

In some sectors, and for rationalized (i.e. efficiency seeking) FI, factors which belong to the generic category of transaction costs and benefits were also of some importance; and particularly those which offered the best opportunities to firms to exploit the economies of scale and scope, and to capture the benefits of the common governance of related activities. In the high technology and knowledge intensive sectors too, MNE affiliates, already well embedded in host countries, were attracted less by low cost labour and more by high quality labour to add value to the O specific assets being transferred from their home countries; and this was particularly so where the affiliates were responsible for complete product lines of their own. These same firms also made stronger demands on the local physical infrastructure, especially transport and communications and frequently valued the presence of sub-national clusters of value-added activities related to their own.

Over the last twenty years, this latter type of FI activity has become increasingly significant, partly because the increasing degree of multinationality of many foreign investors has encouraged subsidiaries to engage in vertical or horizontal specialization, and partly because of the elimination or reduction of distance related transaction costs to FI, especially between countries which are part of a customs union or free trade area, e.g. the European Union. The growth of intra-firm, intra-regional trade has well outpaced that of inter-firm inter-regional trade (UNCTAD–DTCI 1995a, 1996a). It is fashioned by a geographical composition of FI like that of intra-industry trade, in that it is less determined by the country-specific costs of factor endowments or size of local markets, and more by those variables which facilitate firm and/or plant economies of scale and scope,[38] and the effective exploitation of regional and/or global markets.

Most efficiency seeking FI in developing countries tends to be vertically integrated, with investors seeking locations which offer an adequate supply of cost-effective semi-skilled or skilled labor, a good physical infrastructure, government policies which are market friendly, and minimal distance related transaction costs. The external economies of clustering may also offer a locational attraction, particularly in countries whose overall industrial base leaves something to be desired. Such agglomeration economies, first identified by Alfred Marshall (1920), enable the participating firms to draw upon a common infrastructure and a specialized pool of labour or customers, to develop mutually beneficial relations with their suppliers, and learn from local producer associations and their competitors. Hence, the development of export processing or free trade zones, and the deliberate attempts by local or central governments to facilitate market oriented industrial districts of one kind or another.[39] Examples of such industrial districts in developing countries abound.[40] They are particularly numerous in those countries now attracting the bulk of new FI in East Asia – China, South Korea, Malaysia and Indonesia – and in Singapore.

By contrast, most horizontally integrated FI is concentrated in the advanced industrial economies; and it is in these economies – and particularly in some of the knowledge intensive sectors – that one observes not only regional clusters of economic activity, but that FI is especially drawn to these clusters. Again, there are many examples of recently formed clusters.[41] They include, in the US, an agglomeration of biotechnology and semi-conductor firms in the Silicon Valley/Bay area of California, and pharmaceutical and telecommunications firms in New Jersey; in the UK, a range of high technology firms along the M4 corridor (particularly between Slough and Swindon), and financial services in the City of London; and in Japan, the clustering of the primal R&D plants of semi-conductor firms in the Tokyo or Osaka Metropolitan regions (Arita and Fujita, 1996).

Although these districts vary in age, size and character, they, like their counterparts in developing countries, tend to be purpose built – sometimes with the support of local or national governments – and are specifically designed to capture the economies of spatial agglomeration. However, unlike some of their predecessors, e.g. export processing enclaves in developing countries, the business and non-business infrastructure in the new industrial districts is strong and supportive, and their development is linked to the upgrading of indigenous resources, rather than the use of low cost labour or natural resources. There is also considerable collaboration and networking among the participating firms, and between them and non-business and civic institutions, e.g. universities, technical colleges and research and trade associations.[42] Indeed, the evolution of these institutions often follows (and sometimes leads) that of the districts themselves.[43] Frequently, too, firms within the new industrial districts engage in strategic technological and other alliances with those outside the area. This also helps facilitate the learning process of the region; and, indeed, its sustainable competitive advantages (Florida, 1995).

We have discussed, at some length, the changing characteristics of regions and industrial districts within countries, because they illustrate well the changing L characteristics of the geography of FI across national boundaries. In the innovating global economy of the mid-1990s, firms – and particularly MNEs – are increasingly looking for a physical and human infrastructure which enables them to both create and exploit their O specific core competences; and they will tend to prefer locations which best offer the facilities for so doing. Coupled with the need for easy access to external markets, we believe that such a switch of emphasis from traditional cost-minimizing and domestic market related variables explains much of the rising attractions of the most innovating and vibrant industrial or industrializing countries to foreign investors. The obvious exception – which, as we have seen, tends to overwhelm the recent geography of FI – is that of China, and to a lesser extent Indonesia and Thailand, where large domestic markets and low (real) labour costs continue to be the main inducements for inbound MNE activity.

There has, however, been another reaction of firms to recent economic events – and particularly the liberalization of markets and technological advances – which has had a no less critical impact on the geography of FI, particularly among industrial countries. That has been the dramatic increase in the number of cross-border mergers and acquisitions and strategic alliances specifically aimed at protecting or enhancing the global profitability and/or market share of the participating firms. Between 1986 and 1994, for example, it is estimated that world-wide cross-border mergers and acquisitions accounted for between 55 per cent and 60 per cent of FDI outflows; and about three-quarters of those between developed countries (UNCTAD, 1994a, 1995a). Similarly, the number of international strategic alliances in high technology industries concluded annually between 1986 and 1993 was 21.7 per cent higher than between 1981 and 1985.[44]

The spatial implications of FI by firms wishing to sustain or augment their existing O specific advantages, is that they will tend to acquire, or merge with, firms in other locations with a broadly similar, or more sophisticated, structure of natural and created assets to that of their own. Certainly, this applies to mergers and acquisitions and alliances designed to strengthen or complement the technological base of the investing companies, or to accelerate the innovating process. Such asset seeking ventures, together with those aimed at facilitating access to unfamiliar markets, have accounted for a great majority of all cross-border liaisons in recent years (Hagedoorn, 1993, 1996). By contrast, in the newly emerging markets, notably China, wholly or jointly owned greenfield ventures have been the main mode of entry by MNEs. According to UNCTAD (1996a), only one-tenth of cross-border mergers and acquisitions activities take place in developing countries; and, the ratio of mergers and acquisitions to total FI inflows in such countries averaged only 12 per cent to 15 per cent over the period 1988–94. Indeed, one suspects – although there are few hard facts to support this suspicion – there has been more strategic asset seeking FI by Korean, Taiwanese, Malaysian, Thai and Brazilian firms in Europe and the US than there has been of European and US MNEs in Asia or Latin America, and that this has been particularly the case in the first part of the 1990s.[45]

At first glance, the data in Tables 3.1 and 3.2 would appear to lend little support to the idea that MNEs are increasingly engaging in FI to acquire, rather than to exploit, O specific advantages. Even eliminating China from the FI data, the developed countries, and particularly the US, have slightly reduced their share of worldwide investment in the 1990s. What, of course, the data do not, and, by their nature, cannot show is what would have been the geography of FI had not the events of the last twenty years occurred. Our own speculation is that while much of intra-triad merger and acquisition activity has been prompted by technological advances and regional integration in Europe, the engine for FI growth in the rest of the world has been the renaissance of free markets, the introduction of more market friendly government policies, and the 'take-off' of a number of emerging Asian economies. In the absence of the former components of globalization, but not of the latter, the geography of FI might have shifted even more to the developing countries; by contrast, in the absence of the latter, but not the former, we might have seen more concentration of FI in the developed countries. Furthermore, it is worth observing that, like domestic investment by indigenous firms, FI flows are subject to cyclical variations. From 1991 to 1994, FI flows to most developed countries fell from their earlier levels as a result of economic recession; by contrast, FI flows continued to increase throughout the 1990s in both Latin America and Asia, where the rate of growth of gnp exceeded that of the previous decade.[46]

### *Some changes in the modality of organizing the competitive advantages of firms*

Finally, in this section of the chapter, we briefly consider the extent to which the geography of FI may have been affected by changes in the modes by which firms coordinate their cross-border value activities. Notwithstanding the shedding of some activities by firms and the growth of inter-firm collaborative arrangements, FI, as a vehicle for both exploiting and augmenting O specific advantages, was relatively more significant in the 1990–4 period than it was in the 1975–80 period. But, insofar as we believe this to be the case for both developed and developing countries, the impact of the increased I advantages of firms may have had little impact on their geography.[47] For, corresponding to the growth of efficiency seeking FI[48] and asset seeking mergers and acquisitions in developed countries, there has been a much more relaxed attitude to the foreign ownership of domestic assets in other parts of the world, notably in the erstwhile socialist regimes and several Asian and Latin American countries, e.g. Malaysia, Korea, Mexico and Chile. For the most part, in these countries, majority owned FI is now allowed and is accounting for an increasing part of all inbound investment. At the same time, the trend towards more integrated reduction of intra-regional barriers to trade and FI, and production networks – particularly by Japanese and other Asian MNEs in South and East Asia and by US MNEs in Latin America – is further accelerating the pace of FI in these parts of the world.

So much, then, for our suggested reasons for the recent changes in the geography of FI, which we have couched in terms of the OLI paradigm of international production. Table 3.6 summarizes our conclusions on this matter, which, as yet, have not been operationalized or subject to any formal statistical testing.

## Some (home) country, sectoral and firm-specific factors affecting the geography of FI

The impact of global economic events of the last decades on the OLI configuration of MNEs and firms contemplating FI is likely to differ according to their countries of origin, the range and nature of their value added activities, and a gamut of firm-specific structural and strategic-related variables. This means that, *inter alia*, changes in the geography of FI will reflect changes in the value of these contextual variables. For instance, if high value banking and financial services are primarily located in advanced industrial countries, and if FI in such services is growing relative to that in other sectors, then *ceteris paribus*, one would expect the geography of FI as a whole to favour those countries. Developing countries, on the other hand, appear to be the main recipients of FI arising from privatization schemes and a revival of interest by foreign investors in infrastructural development projects (UNCTAD–DTCI, 1996a). Or, in sectors in which plant economies of scale are increasing relative to others, then one would predict an increase in the geographical concentration of FI in those sectors. Finally, to take a firm-specific example, if textile and clothing firms, which currently engage in only labour cost saving activities outside their home countries, are increasing their share of the world

Table 3.6 The changing geography of FI: some suggested explanations couched in terms of the changing OLI configurations facing investing firms

| Advantage | 1975–80 | 1990–4 |
|---|---|---|
| O advantages (of firms) | Those associated with the possession or privileged access to country-specific intangible assets – technology, trademarks, managerial expertise, entrepreneurship; and access to factor, intermediate product of final goods markets<br><br>Prior experience of foreign markets, e.g. by exports, licensing, etc. | Those associated with multinationality *per se* (Ot)<br><br>Organizational expertise and learning experiences and ability to seek out and exploit complementary assets<br><br>Ability to achieve an optimum portfolio of assets and to combine own O advantages with L specific endowments of foreign countries |
| L advantages (of countries) | Traditional L specific variables related to:<br>1 domestic factor costs;<br>2 market-size and growth; and<br>3 transport costs and tariff or other economic and psychic barriers<br><br>Government imposed incentives or obstacles to FI, including performance requirements<br><br>A market facilitating macro-economic and/or macro-organizational policy of host governments<br><br>A stable political and economic regime | The provisions of location-bound resources and capabilities which help firms both to exploit and augment their existing competitive advantages<br><br>The continual upgrading of the assets so as to promote increasingly high-value FI |
| I advantages (of firms) | Those arising from the imperfect markets for specific intangible assets; and from learning experiences and governance of inter-related value-added activities, mainly in the domestic market | Those arising from the ownership and/or control of inter-related activities in different geographical areas. These include spreading of political and environmental risks, and the holistic integration of disparate functions and strategies |

market relative to those which operate a more complete range of value added activities, then one might expect countries which have a revealed comparative advantage in labour intensive sectors will gain from FI relative to those that do not!

Let us now consider some of the ways in which the changing composition of the main capital exporting countries, that of the sectoral pattern of economic activity and that of the characteristics and strategy of the leading MNEs has affected the spatial distribution of FI over the past two decades.

### The structure of source countries

As Table 3.7 shows, the main change in the geographical composition of the stock of outward direct investment between 1973 and early 1995 was a 22.9 percentage point drop in the share of the US, which was matched by a 6.3 percentage point increase in the share of Japan, a 10.9 percentage point increase in that of Western Europe and other developed market economies, and a 6.4 percentage point increase in that of developing countries. Of the other developed countries, most increased their share of outbound FI – and France, Germany, Australia, Spain and Belgium substantially so – but a few, notably the Netherlands, South Africa and Switzerland decreased theirs.

Two questions now arise. The first is: to what extent is the geography of FI of the countries increasing their share of total FI different from that of those decreasing their share? The second is: how far has this influenced the overall geography of FI over the period?

Table 3.8 sets out some details on the geography of FI of Western Europe, Japan and the US[49] for the latter half of the 1970s and the first four years of the 1990s. What do they show? First, in both periods, they depict some major differences in the geography of FI of the three regions or countries. Between 1975 and 1980, Japanese MNEs directed a considerably higher proportion of their investments to developing countries, and especially Asia, than did their European or US counterparts; by contrast, their share of the increased stock of FI in Western Europe was less than one quarter of that of either US or European firms.[50] This geographical pattern, and other differences between the three major investors, e.g. the higher than average share of US FI directed to the rest of the Americas, and that of European FI to other developed countries, e.g. Australia and some ex-colonial territories in Asia, are entirely consistent with the extant explanations of the location of cross-border FI – and, particularly, that of first-time FI.[51]

Second, and perhaps more interestingly, one observes very different changes in the geography of FI of the three major investors since the late 1970s; and with one exception, considerably more so of that of FI taken as a whole. While Japanese MNEs have increased their FI most sharply in the developed countries (and particularly in the US in the 1980s and Europe in the 1990s), US MNEs have reoriented their interests away from other developed countries

*Table 3.7* Stocks of foreign direct investment by major home countries and regions, 1973–95 (billions of US dollars)

| Economy | 1973 | % | 1980 | % | 1990 | % | 1995 | % |
|---|---|---|---|---|---|---|---|---|
| Developed economies | 205.0 | 98.5 | 508.0 | 98.8 | 1,614.6 | 95.9 | 2,514.3 | 92.1 |
| Western Europe | 82.6 | 39.7 | 236.6 | 46.0 | 853.9 | 50.7 | 1,332.5 | 48.8 |
| of which: UK | 27.5 | 13.2 | 80.4 | 15.6 | 230.8 | 13.7 | 319.0 | 11.7 |
| France | 8.8 | 4.2 | 23.6 | 4.6 | 110.1 | 6.5 | 200.9 | 7.4 |
| Germany | 11.9 | 5.7 | 43.1 | 8.4 | 151.6 | 9.0 | 235.0 | 8.6 |
| Italy | 3.2 | 1.5 | 7.3 | 1.4 | 56.1 | 3.3 | 86.7 | 3.2 |
| Netherlands | 15.8 | 7.6 | 42.1 | 8.2 | 109.1 | 6.5 | 158.6 | 5.8 |
| Sweden | 3.0 | 1.4 | 5.6 | 1.1 | 49.5 | 2.9 | 61.6 | 2.3 |
| Switzerland | 7.1 | 3.4 | 21.5 | 4.2 | 65.7 | 3.9 | 108.3 | 4.0 |
| Other | 5.5 | 2.6 | 13.0 | 2.5 | 81.0 | 4.8 | 117.0 | 4.3 |
| North America | 109.1 | 52.4 | 242.8 | 47.2 | 514.1 | 30.5 | 816.0 | 29.9 |
| of which: US | 101.3 | 48.7 | 220.2 | 42.8 | 435.2 | 25.8 | 705.6 | 25.8 |
| Canada | 7.8 | 3.7 | 22.6 | 4.4 | 78.9 | 4.7 | 110.4 | 4.0 |
| Japan | 10.3 | 4.9 | 19.6 | 3.8 | 204.7 | 12.2 | 305.5 | 11.2 |
| Other developed countries | 2.9 | 0.4 | 6.8 | 1.3 | 41.9 | 2.5 | 60.4 | 2.2 |
| of which: Australia | 1.0 | 0.5 | 2.3 | 0.4 | 30.1 | 1.8 | 41.3 | 1.5 |
| Developing economies | 3.0[1] | 1.4 | 6.1 | 1.2 | 69.4 | 4.1 | 214.4 | 7.8 |
| Central and Eastern Europe | neg. | neg. | 0.1 | neg. | 0.2 | neg. | 1.4 | 0.1 |
| Total | 208.1 | 100.0 | 514.2 | 100 | 1,684.1 | 100.0 | 2,730.1 | 100.0 |

*Sources*: UNCTC (1988), UNCTAD (1996a).

*Note*:
1 Calculated by deducting FDI outflows 1974–9 from 1980 stock data (UNCTC, 1988, UNCTAD, 1995a).

(and, in particular, Canada) towards developing countries (and, in particular, Brazil and Mexico in Latin America and Singapore and Hong Kong (but not China) in Asia). European investors, on the other hand, have been increasingly attracted to locations nearer home, especially since the initiation of the Internal Market Program of the European Union (EU) (Dunning, 1997a). Although not documented in Table 3.8, there has also been a reorientation of FI by third-world countries away from other developing countries – the massive increase of FI in China by MNEs from Taiwan, Hong Kong and Singapore is an exception – towards the advanced industrialized countries, and particularly the US and countries in the EU (Dunning, van Hoesel and Narula, 1998).[52]

Third, when comparing the spatial distribution of FI of the three regions with that of all countries, it needs to be appreciated that the relative weight of the three regions as sources of FI has shifted. As Table 3.7 has shown, taken as a group, Western Europe, Japan and the developing countries have increased their share of all FI. Since, in both periods, Japan and the Asian developing countries[53] directed a much greater proportion of their FI to other Asian countries, it is not surprising that the percentage of total FI going to Asian

*Table 3.8* Geographical distribution of changes in outward FDI stock, 1975–80 and FDI flows 1990–4: US, Western Europe and Japan

| Country | 1975–80 | | | | 1990–4 | | | |
|---|---|---|---|---|---|---|---|---|
| | US | Western Europe[1] | Japan | All countries[2] | US | Western Europe[6] | Japan | All countries |
| Developed countries | 78.2 | 81.1 | 45.7 | 76.6 | 60.8 | 82.5 | 68.6 | 74.2 |
| Western Europe of | 48.4 | 40.7 | 9.2 | 43.1 | 49 | 61.8 | 19.5 | 52.4 |
| which: UK | 15.9 | n.a. | n.a. | 16.9 | 18.6 | 7.3 | 7.4 | 10.6 |
| North America | 20.4 | 32.8[3] | 26.9[3] | 27.2 | 6.5 | 17.8 | 43.5 | 17.9 |
| of which: US | (–) | 32.8 | 26.9 | 24.5 | (–) | 16.9 | 41.8 | 15.3 |
| Japan | 3.0 | 0.7 | (–) | 0.5 | 2.3 | 0.1 | (–) | 0.6 |
| Other developed countries | 6.4 | 13.7[4] | 9.6[4] | 5.7 | 3 | 2.9 | 5.6 | 3.3 |
| (of which: Australia and New Zealand) | 3.4 | 10.5 | neg. | 3.9 | 2.7 | 2.5 | 5.5 | 3.0 |
| Developing countries | 21.8 | 18.9 | 53.9 | 23.4 | 37.7 | 14.4 | 31 | 23.5 |
| Asia | 2.8 | 6.3 | 27.7 | 6.2 | 10.3 | 4.5 | 18.8 | 8.1 |
| of which: China | neg. | neg. | neg. | neg. | 0.7 | 0.3 | 8.6 | 1.5[7] |
| Latin America | 14.1 | 7.7 | 15.4 | 12.5 | 25.1 | 5.6 | 10.0 | 11.9 |
| Africa | nsa. | nsa. | nsa. | 2.5 | 0.5 | 1.0 | 1.2 | 0.9 |
| Other | 4.9[5] | 4.8[5] | 10.8[5] | 2.2 | 1.8 | 3.0 | 1.0 | 2.4 |
| Central and Eastern Europe | neg. | 0.1 | 0.3 | neg. | 1.5 | 3.1 | 0.3 | 2.3 |
| Total | 100.0 | 100.0 | 100.0 | 100.0 | 100.0 | 100.0 | 100.0 | 100.0 |
| Share of each region in total FI stock 1980 and 1994 | 42.8 | 46.0 | 3.8 | 100.0 | 25.7 | 49.8 | 11.7 | 100.0 |

*Sources*: 1975–80 – Changes in stocks: UNCTC (1988); 1990–4 FDI flows: OECD (1995).

*Notes*:

1 For UK, Germany and the Netherlands in 1980 accounted for between 70 per cent and 75 per cent of the total Western European outward FDI stock. The UK data are for 1974–80 and the German data are for 1976–80.

2 FDI flows of all countries for both periods. These data are not directly comparable with changes in stock data, and occasionally anomalies occur, e.g. as in the case of the Japanese data, where the US Department of Commerce estimates of changes in stock are considerably in excess of those derived by the IMF from Japanese balance of payments accounts.

3 US data.

4 Including Canada.

5 Including Africa and Unallocated.

6 For UK, Germany, France, Italy, Netherlands and Switzerland, which accounted for 84 per cent of the Western European outward FDI stock in 1994.

7 This figure includes very large investment flows from Hong Kong, Macau, Singapore and Taiwan.

neg. = negligible, n.a. = not applicable.

developing countries – and China in particular – has risen so sharply.[54] On the other hand, as the structure of the Japanese economy has moved closer to that of the leading European economies over the last two decades, so the industrial composition of its FI has changed (Ozawa, 1996); and it has changed in a way which has favoured the advanced industrial countries as a location for both production and pre-production (e.g. R&D) activities. However, the switch of interests of firms from the largest of the three source regions – i.e. Western Europe – from the US to the European Union has more than outweighed the falling attraction of that region to other investors.[55]

We conclude: the data set out in Tables 3.7 and 3.8 suggest that the changing composition of total FI reflects the changing geography of the leading source countries or regions, and their significance as foreign direct investors. Yet, to quite a large extent, these changes have cancelled themselves out. Thus, while Japanese investors have focused more attention on Europe and the US, US investors have been more attracted to the developing economies. Western European investors have continued to favour other industrialized nations, but some of the earlier appeal of the US has diminished, and particularly since 1989. And, while China has emerged as one of the top five recipients of FI since 1990, the greater part of this inflow has originated not from first-world MNEs but from Chinese ethnic communities elsewhere in East Asia.

In our review of recent economic events, we suggested that, apart from first-time foreign investors and those from developing countries, the country of origin of MNEs was becoming a relatively less important source of their core competencies;[56] and that the growth of strategic asset augmenting FI was testimony to this. We also asserted that, the competitive advantages of large MNEs were increasingly arising from their ability to manage and coordinate a network of inter-related cross-border activities – some of which they owned and others which they did not. Such a competency is less likely to reflect the characteristics of the home country of the investing firms, and more the latters' size, organizational structures, and degrees of multinationality and strategies.[57] If our understanding is correct, this would suggest some convergence in the geography of FI is occurring among the leading investing firms. We believe there are some hints of this, in respect of FI in some sectors, and among developed countries.

## The industrial composition of FI

As the previous paragraphs have implied, the geographical patterns of Western European, US and Japanese FI are, at least, partly based on the comparative resource endowments and market conditions of their home countries. Until very recently,[58] most scholars have opined that FI tended to be concentrated in sectors characterized by (one or more of three) features: (1) capital and/or knowledge intensity, (2) product differentiation, and (3) the provision of services which, are supportive of other kinds of FI, are information intensive and/or are 'branded' in some way or another. For much of the postwar period, the growth of both FI and TI has been concentrated in these sectors – notably oil, autos, electronics and electrical equipment, office machinery, pharmaceuticals, packaged foods, banking and finance business consultancies and trade related services – and, indeed, until the late 1980s, the share of the sales of foreign affiliates to the global sales of MNEs in these sectors continued to rise (as did the share of FI in these sectors to all FI). It is, thus, understandable that countries which display a dynamic comparative advantage in those activities are those which have recorded the largest rise in their inbound FI.[59] Such an advantage is partly achieved or advanced by innovating new products; partly by raising the quality, or reducing the cost, of existing products; and partly by the redirection of resources to the production of goods and services which are export or FI intensive. The emergence of Japan, and later some of the more advanced developing countries, as significant outward investors, is explained by their upgraded industrial structures and the need to exploit their unique competitive advantages from a foreign location (Dunning and Narula, 1996).

At the same time, over the last twenty years, a series of critical technological and organizational advances[60] have affected not only the sectoral pattern of economic activity, but the very system within which the production of goods and services takes place. The features of this new production system – variously called 'flexible' 'lean' and 'organcentric' production or Toyotaism – and how it differs from that which it is partly replacing – 'mass' production 'scale' and 'machinocentric' or Fordism – has been the subject of a large number of monographs and scholarly papers.[61] Sufficient for our purposes to observe that these systemic changes, coupled with the growing porosity of national boundaries, the increasingly generic and non-specific nature of many innovations,[62] and the convergence of learning capabilities among networking firms, are tending to blur the distinction between economic activities likely to be trade or FI intensive and those which are not.

So far, these changes, which have been largely concentrated in the advanced industrial economies, have had only a limited affect on the geography of FI. But, comparing the period 1990–4 with that of 1975–80, we find that the extension of the cross-border vertical division of labour has been to embrace higher value activities, and to take advantage of speedier, more efficient and less costly transportation and communication networks. The contemporary textile and clothing industry is an excellent example of a sector which has embraced a wide range of cross-border organizational arrangements, the success of which rests on the application of the latest technological advances, e.g. in computer aided design and manufacturing techniques; in the near-instantaneous transfer of information, e.g. designs, specifications, process technologies and marketing schedules, by use of the internet, fax or e-mail. Other traditional sectors also being upgraded and becoming more FI intensive include building and construction, while the liberalization of markets and privatization schemes and the increasing tradability of many services[63] are resurrecting FI as a delivery mode for a whole range of infrastructural products, notably telecommunications and public utilities, as well as those previously tightly regulated, e.g. banking and financial services, insurance and some professional services.

At this juncture, it is, perhaps, worth observing one important shift in the sectoral composition of FI – which has had considerable spatial implications over the last twenty years. This is the growing significance of intermediate and final services. In 1980, for example, 39 per cent of the total FI stock by US MNEs was in the tertiary sector, and 49 per cent in the manufacturing sector; and the corresponding percentages for (the leading) European and Japanese MNEs were 40 per cent and 37 per cent, and 47 per cent and 34 per cent respectively. By 1993, the percentages of all FI accounted for by the tertiary sector had risen to 51 per cent in the case of the US firms, 49 per cent in the case of European firms, and 66 per cent in the case of Japanese firms; and, that of manufacturing had fallen to 36 per cent, 49 per cent and 27 per cent (OECD, 1996). On average, between 1980 and 1993 services accounted for a little over three-fifths of all new FI; and, if one was also to include the services component of the goods produced in the manufacturing sector – and particularly knowledge-intensive goods – this proportion would probably rise to two-thirds, or even higher in the case of US MNE activity.

Statistical data on the geography of FI in services are extremely patchy. We do know, however, that the strongest growth in MNE activity in services over the last twenty years has been among the triad nations. In the case of the country attracting the largest amount of FI over this period – the US – the share of the new FI stock directed to the tertiary sector was 69 per cent (OECD, 1996). Most of these new capital imports were from Western Europe and Japan; and they comprised both mergers and acquisitions and greenfield investments. Indeed, apart from a few manufacturing sectors, notably autos and consumer electronics and electrical goods, Japanese MNEs have concentrated their foreign activities in financial, trading and transportation services and real estate. Between 1980 and 1993, no less than 70 per cent of the increase in the FI stock of Japanese firms was directed to the tertiary sector (UNCTAD–DTCI, 1993b; OECD, 1996). In the EU too, over the last decade, the share of intra- and extra-EC FI in services – particularly high-value services – has increased quite dramatically.[64] Finally, in the last eight years, there has been substantial cross-border FI in the air-transportation sector.[65]

In many developing countries, too, the growth of inbound tertiary FI has outpaced that of manufacturing FI.[66] However, unlike service related MNE activity in the developed countries (apart from that of first time investors) much of that attracted to the developing world has taken the form of trade enhancing activities, or infrastructural investment, both of which are frequently preludes to FI in the primary or secondary sectors. Thus, for example, since the late 1980s, in China and parts of Latin America FI inflows have been increasingly directed to privatization schemes in such services as electrical power, telecommunications, hotels, and building and construction. Recent deregulation of some public utilities and parts of the financial services sector in India seems set to open the door to more service related FI in that country.

In conclusion, while the implementation of new production systems and the move towards knowledge based capitalism in the advanced industrial economies is reorienting the geography of FI towards these economies, the emergence of powerful new nations in the third world and the liberalization and privatization of the markets for infrastructure and trade supporting services is resulting in a counteracting shift in the location of FI towards the middle and lower income economies. At the moment, the net effect of these forces is favouring the industrialized countries, although one suspects that the current trend could well be reversed if, and when, the third-world countries increase their share of the world's output.

## The changing profile of the leading MNEs

The third contextual variable likely to influence the geography of FI is that of the distinctive characteristics of the MNEs – once one has normalized for sector and country-specific differences. Such characteristics are of two main kinds – which, over time, it may be unwise, let alone very difficult, to distinguish. The first comprise the structural attributes of firms, e.g. their age, size, degree of multinationality, product range, innovating capabilities, degree of vertical integration and so on; and the second comprise their strategic actions or reactions towards the ownership and management of their core competencies, and the location of the value added activities arising from, or associated with, these.

We accept, of course, this is a subject which deserves a separate chapter – perhaps, several chapters! In the paragraphs which follow, we shall confine ourselves to offering a few carefully selected facts about the geography of FI of particular firms, and some speculation about the reasons for them and the extent to which they might help explain the data set out in earlier tables. In our desire to control – as best as we can – for (source) country and sector differences we shall limit our analysis to just six sectors – autos (FI profiles of US, European and Japanese MNEs), food products and petroleum refining (of US and European MNEs), electronics (Japanese and European MNEs), computers (US MNEs) and chemicals (European MNEs).

Table 3.9 sets out the details. It presents a mixed picture. Take, first, the degree of multinationality of these firms in 1994, using an index devised by UNCTAD (1995a).[67] In the petroleum sector, the index varies from 20.0 per cent to 63.8 per cent, but there is almost a perfect correlation between the degree of multinationality and size of enterprise (one of the structural variables earlier identified). The same is true of the three top US auto producers, although the variation in the degree of multinationality is rather less. In the electronics sector, the degree of multinationality, which ranges from 19.3 per cent to 58.5 per cent, does not seem to be size related; rather, in this instance, one suspects it is the distinctive strategy of the firms – and particularly that of the chief executives of the firms towards foreign markets –

Table 3.9 A selection of the largest MNEs classified by sector, country, size[1] and index of multinationality[2] (IM) 1985 and 1994

**Petroleum refining**

| (a) US | | IM 1985 | IM 1994 |
|---|---|---|---|
| Exxon | (3) | 59.0 | 63.8 |
| Mobil | (11) | 57.2 | 58.7 |
| Chevron | (40) | | 30.3 |
| Texaco | (44) | 42.6 | 44.0 |
| Amoco | (64) | 16.8 | 23.3 |
| Atlantic Richfield | (88) | 5.0 | 20.0 |
| (b) European | | | |
| (UK/N) Royal Dutch Shell | (1) | 60.0 | 63.6 |
| (F) Elf Aquitaine | (10) | 43.0 | 56.7 |
| (UK) BP | (27) | 66.3 | 67.2 |
| (I) ENI | (37) | n.a. | 28.1 |
| (F) Total | (73) | n.a. | 68.0 |

**Food products**

| (a) US | | IM 1985 | IM 1994 |
|---|---|---|---|
| Philip Morris | (28) | 28.0 | 41.0 |
| Pepsico | (72) | 29.8 | 38.0 |
| Sarah Lee | (86) | 26.3 | 48.3 |
| RJR Nabisco | (95) | 31.9 | 41.5 |
| (b) European | | | |
| (C) Nestle | (13) | 98.0 | 86.5 |
| (UK/N) Unilever | (23) | n.a. | 84.5 |
| (UK) Grand Metropolitan | (42) | 38.0 | 42.0 |

**Chemicals**

| (a) European | | IM 1985 | IM 1994 |
|---|---|---|---|
| (G) Bayer | (21) | 50.0 | 72.5 |
| (G) Hoechst | (34) | 41.5 | 64.6 |
| (F) Rhône Poulenc | (35) | 32.9 | 61.8 |
| (C) Ciba Geigy | (36) | 66.5 | 64.6 |
| (G) BASF | (45) | 30.2 | 51.5 |
| (B) Solvay | (71) | 89.5 | 92.2 |
| (N) AKZO | (93) | 65.0 | 79.3 |
| (Ny) Norsk Hydro | (99) | 73.6 | 43.5 |

**Computers**

| (a) US | | IM 1985 | IM 1994 |
|---|---|---|---|
| IBM | (5) | 41.5 | 56.4 |
| Hewlett Packard | (59) | 37.1 | 41.4 |
| Digital | (81) | 38.0 | 57.2 |

**Electronics**

| (a) Japanese | | IM 1985 | IM 1994 |
|---|---|---|---|
| Matsushita | (16) | 5.5 | 39.8 |
| Sony | (19) | 20.7 | 58.5 |
| Hitachi | (22) | 15.0 | 27.7 |
| Toshiba | (41) | 6.0 | 20.0 |
| NEC | (57) | 7.5 | 19.3 |

**Autos**

| (a) US | | IM 1985 | IM 1994 |
|---|---|---|---|
| Ford | (2) | 41.0 | 28.6 |
| General Motors | (4) | 30.8 | 25.7 |
| Chrysler | (75) | 16.0 | 15.4 |

| (b) European | | IM 1985 | 1994 |
|---|---|---|---|
| (S/C) ABB. | (15) | 60.0 | 88.4 |
| (F) Alcatel Alsthom | (18) | n.a. | 58.9 |
| (N) Philips | (24) | 93.6 | 85.0 |
| (G) Siemens | (25) | 28.0 | 47.3 |

| (b) European | | IM 1985 | 1994 |
|---|---|---|---|
| (G) Volkswagen | (5) | 29.0 | 60.4 |
| (G) Daimler Benz | (9) | 21.0 | 42.8 |
| (I) Fiat | (20) | 25.0 | 47.0 |
| (F) Renault | (26) | 31.8 | 43.7 |
| (S) Volvo | (39) | 38.0 | 66.6 |
| (G) Bosch | (65) | 25.0 | 54.4 |
| (G) BMW | (66) | 15.0 | 47.2 |
| (c) Japanese | | | |
| Toyota | (8) | 12.5 | 28.1 |
| Nissan | (14) | n.a. | 32.2 |
| Honda | (51) | 21.1 | 41.0 |

*Notes:*
1 Ranked (by size of foreign assets) of the 100 leading MNEs.
2 Defined as a average of percentage of global assets, sales and employment of NMEs accounted for by foreign affiliates.
3 Asea and Brown Boveri.
4 Previously Standard Oil of California.

European country codes
B = Belgium
F = France
G = Germany
I = Italy
N = Netherlands
Ny = Norway
S = Sweden
C = Switzerland
UK = United Kingdom

Above average growth of IM

Below average growth of IM

Average growth of IM

which is the more important explanation.[68] The multinationality index for European firms in both the chemicals and electronics sectors shows a fairly uniform pattern, but here the size of the investing country enters into the picture. In the chemicals sector, Solvay, a Belgian firm and AKZO, a Dutch firm, exhibit well above the average multinationality ratio of the seven firms considered; while, in the electronics sector, Philips, from the Netherlands, and ABB (Asea Brown Boveri, a Swiss/Swedish firm) exhibit very high ratios indeed. More generally, taking country (or in the case of Western Europe, regional) and sectoral considerations into account,[69] the UNCTAD data show a positive correlation between size of firm and degree of multinationality, and, in Western Europe, a negative correlation between size of country and degree of multinationality.[70]

Since the late 1970s, the growing significance of the foreign operations of the world's leading 100 industrial MNEs has largely mirrored that of the countries and sectors of which they are part. That of the Japanese firms has expanded the most (admittedly from a very low base), followed by that of Western European firms – and particularly those from Germany and France. On average, the degree of multinationality of US MNEs has changed very little – and has hovered around the 30 per cent level.

However, within countries and sectors, some firms stand out above others in the aggressiveness with which they pursue their FI strategies. Of those listed in Table 3.9, firms worthy of special mention (their names are lightly shaded) include the (relatively) smaller US oil refining companies Amoco and Atlantic Richfield; Volkswagen, Volvo and BMW among the European auto companies; Hoescht, Rhone Poulenc, BASF and Bayer among the European chemical companies; ABB and Siemens among the European electronics companies; and Matsushita and Sony among the Japanese electronics companies. In other sectors, Digital and IBM (US – computers), Grand Metropolitan (UK – food products) and Sara Lee (US – food products) each recorded increases in their foreign participation ratios. It is, perhaps, worth observing that the growth in these ratios has been accomplished both by purchases of existing foreign firms, e.g. Grand Metropolitan's acquisition of General Mills in the US, and the merger between Asea and Brown Boveri to form ABB; and by aggressive greenfield investments, e.g. those undertaken by IBM, BMW and the Japanese electronics and auto companies in Europe and the US.

By contrast, in the case of those firms shaded more darkly in Table 3.9, the index of multinationality has either declined or not kept pace with the average of the sector of which they are part. As far as MNEs from small countries are concerned, e.g. Ciba Geigy and Nestle (Switzerland), Philips (the Netherlands) and Solvay (Belgium) this is because, the foreign, *vis à vis* the domestic component of their operations, was already very substantial indeed in the early 1980s. In other instances the respective companies have focused attention on improving their position in their domestic markets. For example, faced with intensive competition from their Japanese counterparts, US auto MNEs – and especially Ford and GM – completely refashioned their domestic production systems in the late 1980s and early 1990s, and have since regained much of the sales they lost in earlier years. Domestic mergers and acquisitions have had an ambivalent effect, depending on the geography of the acquired firm's investment.[71] Among the leading 100 MNEs whose index of multinationality has fallen, or has not kept pace with the average for their sector for this reason, are BAT Industries (UK – tobacco, cosmetics and insurance), AT&T (US – telecommunications, equipment and services), Xerox (US – scientific and photographic equipment) and RTR Nabisco (US – food and tobacco products).

What, now, of the consequences of these shifts in the FI profiles of some of the world's largest companies on the geography of FI? Shortage of data constrains us to perform only the most superficial exercise. However, by comparing changes in the geography of the sales, assets or employment[72] of firms according to the extent to which they had increased their index of multinationality in particular sectors and countries, we can gain a hint of the contribution of firm-specific factors to the changing spatial distribution of FI.

We took our data from the 1980 and 1992 editions of the World Directory of MNEs (Stopford, Dunning and Haberich, 1980; Stopford, 1992). We calculated – insofar as the data allowed – the average rates of FI growth for two groups of firms, in each of the main regions of the world. To obtain our groups, we simply divided the total number of firms into two and took the top and bottom growers in each industry/country sub-group.

The results of our exercise were somewhat inconclusive. While there is some suggestion that the MNEs with above average rates of growth grew relatively faster in Europe and North America than MNEs with below average rates of growth, this was not so in the case of developing countries; indeed, in Asia, the slower growing MNEs increased their share of total MNE activity.[73] However, since the number of sample observations was only very small (56), we believe more broader based research is necessary before we can draw any reasonable conclusions about the role of firm-specific factors in influencing the changing geography of FI.

We would make just one other comment about the role of firm specific variables in affecting the geography of FI. Earlier, we stated that, as a result of recent economic and political events and the reactions of governments and firms to them, there was tendency for the O specific advantages of firms to become less dependent on the advantages of their countries of origin, and more on those arising from their multinationality. If this proposition is correct we would expect that variations in the extent of the multinationality of firms would depend less on country of origin and more on firm specific variables such as the size of firms. Table 3.9 sets out data on changes in the degree of multinationality of the largest 100 firms in six of the most internationalized industrial sectors, classified by both country and size. These data suggest that cross-country differences in that index have, indeed, narrowed over the last decade; while there is reason to suppose, that, in some sectors at least, size of firm has become more closely linked with degree of multinationality. There may well be other significant structural or strategic firm specific factors but, at the very least, the data do point to a change in the contextual variables now affecting a firm's geography of FI as compared with those of only a decade ago.

## A note on the role of outward MNE activity in affecting the geography of FI

At a global level, the geography of outward FI should exactly mirror that of inbound FI. We would, however, make two points with respect to the interface between the two kinds of FI. The first is that, depending on the circumstances at the time, the outward FI of any particular region or country can either substitute for, or complement, its DI; but, in both cases, it will almost certainly affect the structure of the latter. Insofar as inbound FI and DI are influenced by similar factors, it follows that one has to relate – at least at a country and industry level – the geography of the former to that of its outward counterpart.

The second point reinforces this view by suggesting that, over the last twenty years, not only has the ratio of inward and outward FI become more balanced in the case of the major foreign investors,[74] but that an increasing proportion of FI flows are now *intra-* rather than *inter*-industry. This suggests that the evolving pattern of FI has tended to follow that of trade – from that based on the spatial distribution of natural resource endowments (i.e. Heckscher–Ohlin type trade), to that based on the economies of scale and product differentiation (Smithian type trade), to that based on the disposition and rate of change of created assets (i.e. Schumpeterian type trade).[75] Initially, FI flowed to developing countries to provide the technology, capital and markets necessary for the production and export of primary products; and to developed countries to exploit their local markets or favoured cost structures, or in response to tariffs or other barriers to exports. Today, the greater part of FI – particularly among the triad countries – is of a Smithian or Schumpterian type, and is designed to capture gains similar to those arising from trade in intermediate products,[76] plus the additional O specific advantages arising from specific economies of scale and scope.

The implication of this phenomenon is that, over time, there is a tendency for the sectoral patterns of inward and outward FI to converge – although this tendency may be halted or reversed as significant new emerging economies, e.g. China and India, enter the world stage as hosts to FI. But, essentially, in the knowledge or scale intensive goods and service sectors, which are frequently dominated by a relatively few major global players, inward and outward FI are complementary to, rather than substitutable for, each other, and there is a great deal of cross hauling of created assets. As a consequence of this, in the short run, at least, the DI of a country may be relatively unaffected by inward and outward FI. We write 'may be' because much depends on the *raison d'être* for MNE activity. If, for example, the L specific advantages of (say) Italy fall relative to those of (say) France, there might be less French, or other FI in Italy, but also more FI in Italy by French and other firms. If, on the other hand, Italian firms improve their efficiency relative to their French counterparts, and the L advantages of the two countries remain the same, there is likely to be an increase in inbound FI to France, while outbound FI will either be unaffected or fall. A third possibility is that, as a result of past inward investment, the O advantages of domestic firms are upgraded. While this will enhance their ability to capture foreign markets by outward FI, the effect on inward FI will be more ambivalent.

These examples serve to illustrate the interdependence between inbound and outbound FI, as well as the critical role governments may play in influencing the level of, and balance between, the two. The conclusion we draw from this analysis is that, to fully understand the rationale of the geographical distribution of FI − particularly among developed countries − the character of that investment and the way it relates to the outbound FI of the recipient country needs to be taken into account. The evidence suggests that between the second half of the 1970s and the first half of the 1990s any increased share of FI directed to developed countries was partly a result of the growth of the cross hauling of intra- rather than inter-industry MNE activity.

## Summary and conclusions

The principal objective of this chapter has been to describe and offer some explanations for the changing geography of FDI between the second half of the 1970s and the first five years of the 1990s. The two periods were chosen to reflect the years immediately preceding the introduction of a series of critical and far reaching economic, political and technological events; and those after the most dramatic of these events − the collapse of the command economic system of the Central and Eastern European economies − came into effect.

The main changes in the geography of the leading investing regions or countries between 1975 and 1980 and between 1990 and 1994 are, first, the emergence of China as the third largest recipient of FI. This surge of inbound FI to the Peoples Republic explains a large majority of the increased share of FI directed to the developing countries since the late 1970s. Second, and more significant, that the changing share of North/South FI flows have been the changes in the distribution of North/North, and North/South flows. Thus, Western Europe − and, in particular, France and Spain − have gained as FI recipients at the expense of Canada and the US; while the NICs of South and East Asia have become more attractive locales relative to most countries in Latin America. Third, while Africa continues to be of marginal interest to foreign investors, Central and Eastern Europe, and especially the Visegràd countries,[77] have begun to emerge as quite important recipients. Fourth, some of the traditional resource based recipients of FI – Canada and Australia – have lost ground to the faster growing industrialized countries. Fifth, inward FI going to Japan has increased only marginally over the last two decades. Sixth, there has been a slight fall in the geographical concentration of FI among developed countries; and a slight rise of that among developing countries. However, excluding China, there has been a sharp reduction in the concentration of FI among developing countries. In our explanations of these facts, we concentrated on the following six factors.

First, a substantial part of the changing geography of FI can be explained by the same variables which explain the changing geography of the TI of countries. However, data on the ratio between FI and TI do suggest that the growth of the former has been especially marked in parts of the European Community, e.g. in France, in Canada and in parts of South and East Asia (especially China) and Central Europe. By contrast, it has been least marked in West Asia and most of Africa and Latin America.

Second, changes in the geography of FI, which may be specifically attributed to its foreign-ness, and/or the degree of multinationality of the investing firms, can usefully be explained by viewing the impact of global political and economic events on the configuration of OLI advantages facing foreign investors or potential foreign investors. After identifying these events, we concluded that both the core competencies of firms and the locational advantages of countries had undergone a number of profound changes (as summarized in Table 3.6); as, indeed, had the organizational modalities by which firms spatially reconfigured the creation and use of their core competencies. In particular, we observed the rising importance of asset acquiring or augmenting FI, particularly among triad countries; and of the need of firms to site their foreign activities in countries and/or regions which offered the quantity and quality of L bound created assets which best complemented their own O specific advantages. These developments have affected the geography of FI in as much as firms increasingly favour locations which offer these facilities – the nature of which will vary with the type of value-added activity undertaken by MNEs, and particularly its technological and knowledge intensity.

Third, as might be expected, recent changes in the geography of FI have reflected the composition of the countries of origin, the nature of the economic activities undertaken, and the structure and strategies of the participating firms. Between 1975 and 1980 and between 1990 and 1994, the most significant growth in outward FI was recorded by MNEs from Japan and from third-world countries (especially from ethnic Chinese communities); by contrast, the US MNEs achieved only a modest growth in their foreign activities. Since, as Table 3.3 shows, the geography of FI by the faster and slower growers was quite different – especially within the developing economies – it is clear that part of the changes in the spatial distribution of FI can be put down to the differential rate of growth of the leading outward investors. A good case in point is that of China – where three-quarters of the inward FI over the past fifteen years has originated from other Asian countries.[78] By contrast, the decreasing share of outward FI accounted for by the US has meant a fall in the share of the total FI directed to Latin America – an area in which US firms tend to invest more than those of MNEs from other countries.

Fourth, we have suggested that changes in the sectoral composition of economic activity have had a more ambiguous affect on the geography of FI. This is because, on the one hand, the growth of the technology and knowledge industrial sectors and high value service sectors has encouraged more FDI in the most advanced economies, while on the other, the liberalization of many infrastructural service sectors, and the need by fast growing emerging economies for resource based and/or market seeking FI has revitalized Heckscher–Ohlin type investment flows. We also stressed some of the (no less ambiguous) implications of the relatively faster growth in services FI for the geography of FI; and also those of the trend towards more flexible production systems, which, we suggested, favoured the growth of small and medium size MNEs, and also the upgrading of both manufacturing and service activities in developing countries.

Fifth, we took a brief look at the implications for the geography of FI of the differential rates of growth of some of the leading MNEs from six industries and three groups of investing countries (or regions). While we could find no satisfactory comprehensive explanation of the changing locational patterns of the faster versus the slower growing MNEs, we were able to conclude that, first, the larger firms of the leading 100 industrial MNEs generally demonstrated a higher index of multinationality than their smaller counterparts; and second, that firms with the lower indices of multinationality in 1980 tended to record higher than average rates of FI growth over the following decade.

Finally, we considered some implications of the convergence in the rates of outward and inward FI of most of the leading developed and developing countries for the geography of FI and concluded that this was helping to fashion a new division of labor based primarily upon the spatial distribution of *created* assets, including those for which governments were, directly or indirectly, responsible. We also suggested that in countries where outward and inward FI were fairly evenly balanced, that the determinants of DI and FI were becoming more difficult to separate from each other. While, too, the contribution of *net* inflows to the total domestic capital formation was often quite small, the *structural* consequences of a sizeable amount of both inbound and outbound FI were frequently very considerable indeed.

In conclusion, we would make a few policy remarks. This chapter has high-lighted the fact that global events of the past two decades – and particularly those associated with technological change and the liberalization of markets – have changed the kinds and determinants of trans-border activities engaged in by firms. The socio-institutional framework of capitalism and the techno-organization of production systems in advanced countries are having a major effect on both the flows of created assets between countries, and on the L bound capabilities of different countries to produce these assets. We have sug-gested elsewhere (Dunning, 1994b) that the effects of these changes may be both centrifugal and centripetal. On the one hand, there are aspects of flexible manufacturing and interactive learning which favour the resources and capabil-ities of developing countries;[79] and on the other, the need for a physically close and ongoing relationship between firms in dynamic sectors along the value chain are centripetal in their affects. It seems likely that the developing econo-mies which can reconcile these two forces – *inter alia* by a combination of (1) fostering of sub-national and/or regional clusters of inter-related activities, (2) ensuring that there is a strong nucleus of flagship indigenous firms[80] – par-ticularly in internationally oriented sectors – and (3) facilitating infrastructural development and the entrepreneurial and learning talents of their people, are those whose emerging geography of FI is most likely to mirror the efficient tap-ping into, and use of, the world's rich sources of wealth creating assets.

Throughout the world, the emergence of a market and innovation driven globalizing economy, and that of alliance capitalism, are compelling govern-ments, at all levels, to reconsider and reconfigure their macro-economic and macro-organizational policies. Increasingly, too, governments, both of devel-oped and developing countries, are competing with one another for mobile investment. Increasingly, too, as they seek to upgrade the competitiveness of their indigenous resources and capabilities, the more critical their role in the provision of 'sticky', i.e. location bound, assets needed by both home based and foreign based MNEs becomes. Increasingly, too, national govern-ments – particularly in regions which MNEs would like to consider as a sin-gle market – need to work together in order to ensure that firms can fully exploit the benefits of large scale production, and that intra-regional trans-port and transaction costs are kept to a minimum. Increasingly, too, govern-ments may need to harmonize at least some of their policies – notably with respect to FDI regulations incentives and competition – if distorted and coun-ter-productive economic signals are not to be sent out to potential investors. Increasingly, too, technology and organizational capabilities are becoming more generic (i.e. less activity specific). At the same time, to engage in a par-ticular economic activity, firms increasingly need to draw upon a number of specialized or idiosyncratic technologies, and unique skills and learning experiences of other firms; while governments may need to pay more atten-tion to fostering clusters or agglomerations of related activities – and particu-larly those which generate important extra-market or social economies.

In the industrialized world, the benefits of a refashioning of the role of national governments are beginning to be acknowledged, although the speed at which they are being implemented vary considerably, cf. Japan with the US. At a micro-regional level, several US states and provinces, some German länder and same districts in the UK, Italy and Japan are pursuing active policies to attract and retain mobile investment. At a macro-regional level, the EU, and to a lesser extent NAFTA and ASEAN, are excellent examples of ways in which regional authorities may promote or inhibit the networking of firms within their areas of jurisdiction.

## Acknowledgement

I am indebted to Cliff Wymbs of Rutgers University for statistical assistance in the preparation of this chapter.

## Notes

1  As summarized in Dunning (1993) p. 16.
2  All data are expressed in current prices.
3  The annual average growth rate of FDI flows between 1980 and 1994 was 15.1 per cent and for exports of goods and non-factor services 6.4 per cent.
4  It is estimated by UNCTAD, (1996a) that, in 1993, the value of the global sales of the foreign affiliates of MNEs was $6.0 billion, as compared with the global exports of goods and non-factor services of $4.7 billion.
5  The justification for using this measurement is set out in the next section of the chapter.
6  The OLI configuration, explaining the extent and pattern of the foreign value-added activities was first put forward by the author in the mid-1970s. For a recent exposition of the eclectic paradigm of international production, see Dunning (1995a).
7  Of course, we accept that there are other modalities than FDI in promoting the cross-border mobility of goods, services and assets. Indeed, it is likely – though this is very difficult to quantify – that non-equity strategic alliances and networking have become increasingly important vehicles for the transfer of assets, particularly intangible assets, over the past two decades. See particularly, in this connection, the work of John Hagedoorn and his colleagues at MERIT and the University of Limburg.
8  These latter we have termed strategic asset acquiring advantages (Dunning, 1993). In the last decade, FI of this kind – particularly that by firms investing in advanced industrialized countries by way of mergers and acquisitions – has become one of the dominant factors affecting the geography of FI.
9  For further details, see pp. 56–7, and Kenney and Florida (1993, 1994).
10  For example, an FI outflow may lower TI and thus increase the ratio of an FI inflow to TI.
11  An arithmetical example may help to illustrate this point. Suppose in a particular country in year $t$ there is no inward or outward FI and that domestic investment is $100 million. Now suppose in year $t + 1$ there is $20 million inward and $15 million outward FI. Consider two scenarios. The first is that domestic investment by indigenous firms is unaffected by the FI inflows and outflows. In this case, TI increases to $220 million and the FI inflow to TI ratio is $20 million/

$220 million, i.e. 9.1 per cent. The second scenario is where inward FI replaces domestic investment but outward FI reduces domestic investment by $15 million. In this case TI is reduced to $185 million and the FI inflow to TI ratio rises to 10.8 per cent.

12 There have been several attempts by scholars to do this. These are summarized in Chapters 10 and 14 of Dunning (1993), and in Chapter 2 of Caves (1996). The general consensus of these studies is that, where domestic resources are currently underutilized, inbound FI will increase TI, and that, depending on the nature of outbound FI, it may either increase or lower TI. Where resources are fully employed, the net effect on TI will depend on the relative capital intensity of FI, versus TI, and the efficiency at which the former is deployed relative to the latter.

13 We chose the former period as this was immediately prior to the wave of liberalizing markets, and the current generation of technological advances.

14 We fully recognize the limitations of data on FI (FDI) flows, and, most noticeably, the fact that reinvested profits are included in the reporting procedures of some countries, but not in others. We accept, too, that FI is a a form of financing capital expenditure, rather than the expenditure itself, and is obtained from balance of payments data.

15 Domestic investment includes the inbound investment by foreign firms but excludes the foreign investment by domestic firms. Of the various indices of domestic investment which are readily available for most countries, gross fixed capital formation is the one which is the nearest counterpart to FDI.

16 See, for example, IMF (various issues), UNCTAD (1996a).

17 We have taken the composition of the EU to be that existing on 1 January 1995 for all years discussed in this chapter.

18 The US, in particular, substantially increased its share of world wide FI in the 1980s. Between 1983 and 1989 its accounted for 42.6 per cent of all inflows, compared with 24.5 per cent in 1975–80. However, between 1990 and 1994, its share fell back to 10.9 per cent.

19 In the case of both Africa and West Asia, it was the oil exporting countries which recorded the least gains, and, indeed, in the 1990s, the flow of new investment has been less than one-half that of the second half of the 1980s.

20 However, of the larger countries in the world, the one which attracted the least inbound FI until very recently was India. Between 1975 and 1985, the average annual inflow of FI was only $41 million. The interest of foreign investors in India has risen sharply over the last decade, and in the years 1993–5 the rate of increase of annual average inflow was 65.5 per cent.

21 Notably Israel, Greece, Portugal, Sweden, Jamaica and Morocco.

22 Although, as mentioned earlier, FI is a financing, rather than an expenditure, measure.

23 The formula for doing so is set out in the notes to Table 3.3.

24 As set out in Appendix 3.A2 in the form of the ratios of FI inflows to GFCF.

25 However, within these regions, there were large differences. Among the countries with the highest FI/TI ratios were: the UK, Belgium, Ireland, the Netherlands and Greece in Western Europe; Australia in other developed economies; Canada in North America; Malaysia, Fiji, Singapore in Asia and the Pacific; Uruguay, Trinidad and Tobago, Guatemala in Latin America; and Botswana, Egypt, Tunisia, Kenya and Swaziland stand out. By contrast, the Scandinavian countries and Germany in Europe; China, India and Korea in Asia; Jamaica, Venezuela and Ecuador in Latin America; and Algeria, Ghana, Nigeria and Uganda all recorded well below FI/TI ratios. See also Appendix Table 3.A2.

26 The importance of China in the revived growth of FI in developing countries cannot be over-stressed, even though much of this investment originated from overseas Chinese ethnic communities. Excluding that country, the share of FI received by developing countries in the period 1990–4 was 21.3 per cent, marginally *less* than it was between 1975 and 1980 (23.4 per cent).

27 We have selected only countries with annual FI flows averaging more than $50 million over the 1990–5 period. Details of the FI/GFCF ratios for all countries may be calculated using data in various UNCTC (or UNCTAD) and UN (National Accounts Statistics) publications. See especially UNCTC (1988) and UNCTAD (1996a).

28 We would emphasize these are data on *shares* of FI to GFCF. They tell us nothing about the changes in the amounts of either FI or TI.

29 These strategies include whether a particular set of O advantages are exploited via FDI (i.e. the markets for them are internalized (I), or via non-equity cooperative ventures of one kind or another.

30 The kind of O advantages which scholars, in the 1970s, used to explain FI are set out in Dunning (1993) and Caves (1996). The ones which consistently offered the greatest explanatory power were proprietary knowledge and product differentiation and marketing advantages. For a more extensive examination of the role of country specific variables in affecting the O specific advantages of firms see Dunning (1990).

31 Particularly in its subcontracting policies of (relatively) labour intensive products or production processes.

32 In 1973, 89.4 per cent of the FDI stock was accounted for by MNEs from North America and Western Europe. By 1994, the proportion had fallen to 79.9 per cent (UNCTC 1988; UNCTAD 1996a).

33 In 1977, the foreign affiliates of US firms accounted for 8.6 per cent of the world wide R&D undertaken by them; while Japanese firms undertook hardly any R&D outside their home countries. In the period 1973–7, according to Cantwell and Hodson (1991), only 10.4 per cent of the patents registered in the US by the world's largest firms could be attributable to research undertaken in foreign locations (i.e. outside their home countries).

34 In 1993, US firms conducted 13.0 per cent of their R&D outside their home countries; the Japanese were increasingly establishing R&D facilities in Europe; while in the US, foreign firms accounted for 15.5 per cent of total R&D expenditures in 1991, compared with 4.8 per cent in 1977 (Dunning and Narula, 1995).

35 Mainly in respect of high technology FI in the US, and that by US MNEs in Europe and Japan and that by Japanese MNEs in Europe and the US. There is also some suggestion that within the triad countries there has been increased concentration of research related FI in particular regions or districts. See, for example, Dunning and Narula (1995), Kenney and Florida (1993, 1994) and Kuemmerle (1996).

36 Part of the reason for this is that part – and probably an important part – of the growth of foreign based R&D has taken the form of acquisition rather than of greenfield R&D. In the US, almost four-fifths of the total investment outlays by foreign direct investors in the 1980s was through the purchase of existing US businesses.

37 We accept there are certain parallels between the kind of FI designed to acquire natural resources in the nineteenth and early twentieth centuries, and that designed to acquire created assets, notably technology, information and learning experience of the 1980s and 1990s. Both were (or are) aimed at facilitating, or enhancing, the use of the existing O advantages of the investing companies;

and, both were (or are) frequently prompted by aggressive, or defensive, production, marketing and innovatory strategies of large oligopolists.

38 For a review of modern trade theory, see Krugman (1990); for an examination of the evolving relationship between modern trade and FDI theory, see Dunning (1995b).

39 For a description of the various kinds of industrial districts, see Harrison (1992), Gray, Golog and Markusen (1995), Park and Markusen (1995), Enright (1994) and Saxenian (1994).

40 See, for example, Park and Markusen (general) (1992), Park and Markusen (Korea) (1995), Ohmae (China) (1995) and Balasubramanyam (India) (1996).

41 Industrial clusters are, of course, not confined to high technology firms. Examples of the concentration of more traditional industries are to be found in Porter (1990), Enright (1994) and Harrison (1994). However, one's sense is that, in the contemporary global economy, these clusters are becoming more, rather than less, important.

42 Frequently, on the lines set out by Rugman and D'Cruz (1995), in their five partners business network model. The five partners identified by the authors are a flagship or group of flagship firms, their competitors, their key suppliers, their key competitors and non-business infrastructure.

43 And, particularly so in the case of science parks, e.g. in Cambridge, England and Ibaragi and Tsukuba Science City in the Prefecture of Japan, some high value and specialized research, e.g. biotechnology in California.

44 I.e. from 258 to 314. These figures were provided by John Hagedoorn to the author. Also see Hagedoorn (1993, 1996).

45 Such a view would seem to be supported by a very large upsurge of outward FI from these countries to the more advanced industrialized countries since 1990. However, it remains true that only a relatively small proportion of that investment took the form of mergers and acquisitions.

46 In high income countries, the gross domestic product (GDP) rose by 1.7 per cent per annum between 1990 and 1994, compared with 3.2 per cent between 1980 and 1990. The corresponding percentages for Latin America were 3.6 per cent and 1.7 per cent and for East Asia and the Pacific 9.4 per cent and 7.4 per cent (World Bank, 1996).

47 We would emphasize that our claim that FI continues to be a major mode for exploiting or acquiring the O specific advantages of firms in no way negates the proposition that firms need to engage in more cooperative ventures in order to best protect or enhance these advantages. In practice, however, many of these cooperative ventures will be between institutions *within* particular countries rather than *across* countries.

48 Efficiency seeking FI usually implies some kind of integration of cross-border production and markets, over which investing firms prefer to have as much control as possible.

49 These data are culled mainly from OECD (1996), UNCTC (1988) and UNCTAD (1995a). In turn, these data were initially obtained from those published by the IMF (various issues), based on balance of payments statistics and those directly provided by national authorities of FI stocks and/or flows, as revealed by Table 3.7 in 1980, the countries or regions identified accounted for 97 per cent of the FDI stock in 1980 – 94 per cent in 1993.

50 Unlike its US and Japanese counterparts, Western Europe FI *includes* intra-regional FI – which, as the table reveals, has increased substantially over the last twenty years. Were such FI *excluded* from the data, its pattern of FI would look very different indeed.

51  In terms, e.g. of cultural familiarity, low physical and psychic distance costs, political and language ties, similarity of legal and commercial systems, etc., see, for example, the literature reviewed in Dunning (1993) chapters 6 and 7.

52  Of fourteen developing countries for which data are available, ten – China, Hong Kong, India, Indonesia, Korea, Singapore, Taiwan and Thailand in Asia, and Colombia and Mexico in Latin America, directed a higher proportion of their FI stock to developed countries in the early 1990s than in 1980; while MNEs from fourth countries, the Philippines in Asia and Argentina, Brazil and Columbia in Latin America, directed a higher proportion of their FIs to other developing countries. As a whole, an unweighted average of the FI stock allocated to developing countries in 1980 was 63.3 per cent, and to developed countries 36.7 per cent. By 1992, the respective proportions were 56.4 per cent and 43.6 per cent (Dunning, van Hoesel and Narula, 1998; UNCTAD, 1996b).

53  Which, in 1993, accounted for 84.5 per cent of the estimated stock of European investing to outward direct investment in 1993 (UNCTAD, 1995a).

54  About two-thirds of the inbound stock of FI in China in 1994 was accounted for by other ethnic Chinese communities (UNCTAD, 1996a). However, as the transfer of labour intensive production slows down, it is expected that this ratio will fall in the second half of the 1990s.

55  The share of the US European FI reached its peak in the 1986–89 period, when it accounted for 44 per cent of the total FI flows of the four leading investing countries in Europe – UK, Germany, France and Netherlands (OECD, 1996).

56  The concept of the widening geographical source of the O specific advantages of firms is currently being explored in the literature; see, for example, Dunning (1996b).

57  Elsewhere, Dunning (1997b), we have argued that the degree of multinationality is becoming a more important differentiator of types of FI than that of country of ownership.

58  We write 'very recently' as it has only been in the last five years that substantial MNE activity has occurred in sectors in which there was previously no, or little, foreign investment. For an examination of the recent surge in FI in infrastructure development, see UNCTAD (1996a).

59  Assuming that it was desirable to exploit these advantages by way of FI rather than trade or licensing.

60  Or what Carlota Perez (1983) prefers to call changes in 'technological styles'.

61  We will not even attempt to catalogue these, but one very recent analysis of the interface between the new system and the globalization of economic activity is that by Ruigrok and van Tulder (1995).

62  By which we mean that they are not activity specific – at least in their application. Current developments in robotics, biotechnology and telematics have far-reaching implications for a wide range of industrial sectors, with the result that widely differing activities are not only making more use of created (as opposed to natural) factor inputs, but are converging in their use of them

63  The increasing tradeability of services is a two-edged sword as far as FI is concerned. On the one hand, it opens doors for trade in services previously closed; on the other, it facilitates the kind of FI which, itself, makes for more intra-country (and intra-firm) trade in services.

64  In 1984–6, for example, business services – including management consultancy and computer services – accounted for only 0.6 per cent of all FI flows in the EC; by 1990–2, the percentage had risen to 9.9 per cent (Dunning, 1997a).

65  For details, see UNCTAD (1993a and 1995a). It should be noted, however, that much of this FI has taken the form of a minority equity stake.

66 Although, as a proportion of total FI, it is still below that of most developed countries. In 1993, for example, 33 per cent of the stock of inbound FI in Taiwan was in the tertiary sector (compared with 20 per cent in 1980 in Korea. The corresponding percentages were 37 per cent and 21 per cent (UNCTAD, 1995a).

67 This essentially represents an average of three measures of multi- or trans-nationality – share of foreign sales to total sales, share of foreign assets to total assets, and share of total employment. This index was calculated for the top 100 MNEs in 1994, ranked by their foreign assets.

68 Sony, for example, has always been the leader in penetrating foreign markets, with Hitachi and Toshiba lagging behind. The low index for Hitachi reflects the fact that its main products are more knowledge intensive and higher up the value chain.

69 Including the fact that Japan is a latecomer on the multinational scene.

70 This proposition also appears to hold good for the computer sector in the US and the pharmaceutical and food sectors in Western Europe. However, in the auto sector of Japan – mainly because of the distinctive FI strategy of Honda, which paralleled that of Sony in electronics – there was no such correlation parallel.

71 It is possible, too, that cross-border mergers and acquisitions can lead to a fall in the index of multinationality. One example would be where the acquired firm has more than half its activities located in the acquiring country.

72 Depending on the information published by firms.

73 Thus, for example, the average ratio between the 1990 and the 1978 share of the sales (and/or employment) of foreign affiliates in Europe, relative to all foreign sales of MNEs in the top half of the industry/country groupings, was 1.15; and for MNEs with below average it was 1.05; corresponding ratios for North America were 1.97 and 1.31. For developing countries, the ratios for Asia were 0.94 and 1.35; and for Latin America 0.93 and 0.90. It will be observed that the time period chosen for this exercise was that in which the share of all FI in the US increased sharply.

74 Thus, for example, the ratio of the outward to inward FDI stock for the US in 1980 was 2.65; in 1993 it was 1.26. Corresponding ratios for the UK were 1.28 and 1.29; Germany – 1.18 and 1.45; the Netherlands – 2.20 and 1.54; Spain −0.23 and 0.25; Canada −0.42 and 0.82; Australia −0.17 and 0.38; Korea 0.12 and 0.50; Malaysia −0.07 and 0.12; Taiwan −0.04 and 1.47; Singapore −0.10 and 0.12; Brazil 0.04 and 0.12; and Mexico −0.015 and 0.025.

75 The distribution between these three forms of trade is made by several scholars – most recently by Dunning, van Hoesel and Narula (1998).

76 Indeed, the MNE is, itself, an excellent example of a creator of Schumpterian type trade, as its very *raison d'être* lies in the export of created assets to its foreign affiliates; while efficiency seeking FI is specifically designed to capture the advantages arising from both plant economies of scale and firm economies of horizontal and/or vertical integration (Markusen, 1995).

77 That is, the Czech Republic, Hungary and Poland. These countries accounted for 69 per cent of the regions stock on inbound FI in 1994 (UNCTAD, 1995a: 100a).

78 Indeed, paralleling China's growing share of inward investment has been Hong Kong's increasing share of outward investment. Whereas in 1980, Hong Kong accounted for 0.03 per cent of the stock of all outward FI and 2.4 per cent of that of all developing countries, in 1995 the corresponding proportions were 3.1 per cent and 39.7 per cent (UNCTAD, 1996a).

79 Because, *inter-alia*, of the increasing value of individual entrepreneurship and organizational competence as competitive advantages, and because the optimal plant size of many value added activities is falling. Even in the advanced indus-

trial economies, one is witnessing a resurgence of the role of the small and medium size firms in the innovating process.

80 Sometimes, e.g. as in the case of smaller economies such as Singapore, these indigenous firms may be state owned and/or joint ventures with foreign firms.

# APPENDIX 3.A1

## The growth of inward FDI flows, 1975–80 and 1990–5

| | 1975–80[1] $m | 1990–5[1] $m | Growth % 1975–80=100 | | 1975–80[1] $m | 1990–5[1] $m | Growth % 1975–80=100 |
|---|---|---|---|---|---|---|---|
| **Developed countries** | | | | | | | |
| *Western Europe* | | | | *Latin America and the Caribbean* | | | |
| Austria | 139 | 837 | 6.02 | Argentina | 300 | 3,299 | 11.00 |
| Belgium/Lux | 1,203 | 8,336 | 6.93 | Brazil | 1,835 | 2,229 | 1.21 |
| Denmark | 75 | 2,297 | 30.62 | Chile | 118 | 1,365 | 11.57 |
| Finland | 44 | 783 | 17.79 | Columbia | 86 | 928 | 10.79 |
| France | 2,127 | 18,031 | 8.48 | Ecuador | 49 | 310 | 6.33 |
| Germany | 1,052 | 2,568 | 2.44 | Paraguay | 29 | 131 | 4.51 |
| Greece | 404 | 1,022 | 2.53 | Peru | 110 | 629 | 5.72 |
| Ireland | 244 | 94 | 0.39 | Venezuela | −44 | 729 | −16.57 |
| Italy | 553 | 3,702 | 6.69 | Bermuda | 449 | 2,569 | 5.72 |
| Netherlands | 1,277 | 7,843 | 6.14 | Costa Rica | 56 | 194 | 3.47 |
| Portugal | 90 | 1,848 | 20.53 | Dominican Republic | 61 | 180 | 2.95 |
| Spain | 970 | 10,918 | 11.26 | Honduras | 11 | 54 | 4.91 |
| Sweden | 100 | 5,343 | 53.43 | Jamaica | −6 | 127 | −21.17 |
| United Kingdom | 5,195 | 19,673 | 3.79 | Mexico | 1,023 | 5,172 | 5.06 |
| Gibraltar | n.a. | 73 | – | Nicaragua | 7 | 29 | 4.14 |
| Norway | 383 | 761 | 1.99 | Saint Vincent & Gren. | n.a. | 22 | – |
| Switzerland | n.a. | 2,711 | – | Trinidad & Tobago | 119 | 271 | 2.27 |
| | | | | Virgin Islands | n.a. | 165 | – |
| *North America* | | | | | | | |
| Canada | 862 | 6,222 | 7.22 | *Europe and Middle East* | | | |
| United States | 7,895 | 39,774 | 5.04 | Malta | 19 | 56 | 2.95 |
| | | | | Cyprus | 51 | 90 | 1.76 |
| *Other developed countries* | | | | Oman | 90 | 126 | 1.40 |
| Australia | 1,271 | 6,182 | 4.86 | Saudi Arabia | n.a. | 924 | – |
| Israel | 21 | 416 | 19.81 | Syrian Arab Rep. | n.a. | 70 | – |
| Japan | 152 | 1,259 | 8.28 | Turkey | 46 | 769 | 16.72 |
| New Zealand | 229 | 1,992 | 8.70 | United Arab Emirates | 63 | 74 | 1.17 |
| South Africa | −90 | −3 | 0.03 | | | | |
| | | | | *Asia and the Pacific* | | | |
| **Developing countries** | | | | Cambodia | n.a. | 59 | – |
| *Africa* | | | | China | n.a. | 19,635 | – |
| Egypt | 376 | 699 | 1.86 | Hong Kong | 241 | 1,680 | 6.97 |
| Libya | −659 | 133 | −0.20 | India | 41 | 516 | 12.59 |
| Morocco | n.a. | 445 | – | Indonesia | 289 | 1,913 | 6.62 |
| Tunisia | 104 | 209 | 2.01 | Korea | 61 | 931 | 15.26 |
| Angola | 7 | 278 | 39.71 | Lao | n.a. | 36 | – |
| Cameroon | 47 | 19 | 0.4 | Malaysia | 524 | 4,444 | 8.48 |
| Ghana | 16 | 110 | 6.88 | Pakistan | 33 | 74 | 2.24 |
| Kenya | 53 | 18 | 0.34 | Philippines | 74 | 880 | 11.89 |
| Nigeria | 163 | 1,140 | 6.99 | Singapore | 502 | 4,780 | 9.52 |
| Swaziland | 24 | 56 | 2.33 | Sri Lanka | 15 | 128 | 8.53 |
| Zambia | 37 | 99 | 2.68 | Taiwan | 91 | 1,207 | 13.26 |

## Appendix 3.A1 *(cont.)*

| | | | |
|---|---|---|---|
| Thailand | 85 | 1,873 | 22.04 |
| Viet Nam | na | 58 | – |
| | | | |
| *Central and Eastern Europe* | | | |
| Albania | n.a. | 40 | – |
| Bulgeria | n.a. | 66 | – |
| Czech Republic | n.a. | 1,310 | – |
| Estonia | n.a. | 161 | – |
| Hungary | n.a. | 1,987 | – |
| Latvia | n.a. | 134 | – |
| Poland | n.a. | 1,193 | – |
| Romania | n.a. | 185 | – |
| Russian Rep. | n.a. | 1,327 | – |
| Slovakia | n.a. | 217 | – |
| Ukraine | n.a. | 190 | – |
| Kazakhstan | n.a. | 180 | – |
| Uzbekistan | n.a. | 62 | – |
| Croatia | n.a. | 86 | – |
| Slovenia | n.a. | 114 | – |

*Sources*: UNCTC, 1988
UNCTAD, 1996a.
*Note*:
1 = Annual average.

# APPENDIX 3.A2

## Changing share of FDI inflows to gross domestic capital formation, 1976–80 and 1990–4

| | Annual average (%) 1976–80 | 1990–4 | | Annual average (%) 1976–80 | 1990–4 |
|---|---|---|---|---|---|
| Developed countries | | | Developing countries (cont.) | | |
| *Western Europe* | | | *Latin America and the Caribbean* | | |
| Austria | 0.9 | 1.8 | Argentina | 2.1 | 29.3 |
| Belgium/Lux | 5.8 | 22.3 | Brazil | 3.9 | 3.2 |
| Denmark | 0.3 | 8.7 | Chile | 4.2 | 10.0 |
| Finland | 0.4 | 4.2 | Columbia | 2.2 | 11.0 |
| France | 1.9 | 7.1 | Ecuador | 1.8 | 9.2 |
| Germany | 0.8 | 0.4 | Paraguay | 3.9 | 7.9 |
| Greece | 5.4 | 7.9 | Peru | 2.0 | 7.8 |
| Ireland | 6.7 | 1.2 | Venezuela | −0.9 | 8.2 |
| Italy | 0.8 | 1.5 | Costa Rica | 6.0 | 12.8 |
| Netherlands | 4.5 | 12.2 | Dominican Republic | 5.0 | 8.4 |
| Portugal | 1.5 | 10.7 | Honduras | n.a. | 6.6 |
| Spain | 2.8 | 10.1 | Jamaica | 1.6 | 11.7 |
| Sweden | 0.5 | 11.1 | Mexico | 3.6 | 7.4 |
| United Kingdom | 8.4 | 10.5 | Nicaragua | n.a. | 5.2 |
| Gibraltar | | | Saint Vincent & Gren. | n.a. | 39.7 |
| Norway | 3.2 | 3.3 | Trinidad & Tobago | 10.7 | 42.2 |
| Switzerland | n.a. | 4.8 | | | |
| | | | *Europe and Middle East* | | |
| *North America* | | | Oman | 8.9 | 7.0 |
| Canada | 1.7 | 4.8 | Syrian Arab Rep. | n.a. | 1.1 |
| United States | 2.0 | 4.1 | Turkey | n.a. | 1.9 |
| | | | United Arab Emirates | n.a. | 0.8 |
| *Other developed countries* | | | | | |
| Australia | 4.6 | 7.6 | *Asia and the Pacific* | | |
| Israel | 1.1 | 2.7 | Cambodia | n.a. | n.a. |
| Japan | 0.1 | 0.2 | China | 0.1 | 11.6 |
| New Zealand | 6.1 | 24.2 | Hong Kong | 4.2 | 6.7 |
| South Africa | | | India | 0.1 | 0.5 |
| | | | Indonesia | 2.4 | 3.5 |
| Developing countries | | | Korea | n.a. | 0.7 |
| *Africa* | | | Lao | n.a. | n.a. |
| Egypt | 7.1 | 6.7 | Malaysia | 11.9 | 22.4 |
| Libya | n.a. | 1.6 | Pakistan | 0.9 | 3.6 |
| Morocco | 1.6 | 7.0 | Philippines | 0.9 | 6.2 |
| Tunisia | 6.1 | 7.0 | Singapore | 16.6 | 28.4 |
| Angola | n.a. | 31.7 | Sri Lanka | 2.6 | 4.6 |
| Cameroon | n.a. | 1.7 | Taiwan | 1.2 | 3.0 |
| Ghana | 1.8 | 7.7 | Thailand | 1.5 | 2.9 |
| Kenya | 4.4 | 1.1 | Viet Nam | | |
| Nigeria | 0.5 | 29.7 | | | |
| Swaziland | 18.8 | 33.9 | | | |
| Zambia | n.a. | 21.8 | | | |

Appendix 3.A2 (*cont.*)

| *Central and Eastern Europe* | | |
|---|---|---|
| Albania | | |
| Bulgeria | n.a. | 3.0 |
| Czech Republic | n.a. | 8.5 |
| Estonia | n.a. | 43.0 |
| Hungary | n.a. | 22.0 |
| Latvia | n.a. | n.a. |
| Poland | n.a. | 6.8 |
| Romania | n.a. | 4.1 |
| Russian Rep. | n.a. | 0.1 |
| Slovakia | n.a. | 8.0 |
| Ukraine | n.a. | n.a. |
| Croatia | n.a. | n.a. |
| Kazakhstan | n.a. | n.a. |
| Uzbekistan | n.a. | n.a. |

*Sources*: UNCTC, 1988
UNCTAD, 1996a.

# Part II

# CONVENTIONAL SOURCES OF FDI AND TECHNOLOGY

This part presents more detailed case studies of FDI and technology outflows originating in conventional source countries of FDI and technology, namely the triad nations. As pointed out earlier, the geography of FDI and hence of technology transfer is significantly affected by the country composition of outflows, so a more detailed look at the patterns emerging in outflows from triad countries would be useful. Three major sources of FDI flows, namely the US, Germany and Japan, have been selected for detailed examination. The bulk of the European FDI outflows are absorbed within the other regional member states of the European Union and a considerable and growing proportion of the outflows is directed to the East and Central European countries. Germany's case is typical from that point of view and hence is selected from among the major European outward investors to examine the implications of the geographical shift in German MNEs' focus. Japanese FDI outflows also reveal a strong regional focus and hence special attention is paid to their investments in the Southeast and East Asian developing countries. Chapter 4 examines the trends in the internationalization of the US corporations over the 1977–94 period and prospects for developing countries. Chapter 5 examines the recent growth of German FDI outflows, the patterns of its regional distribution and the prospects of diversion of German investments away from developing countries in future. Chapter 6 deals with Japanese FDI especially in the context of Asian developing countries.

# 4

# THE INTERNATIONALIZATION OF US MNEs AND ITS IMPACT IN DEVELOPING COUNTRIES

*Robert E. Lipsey*

## Introduction

There are many possible definitions of internationalization, some referring to the whole economy of the home, or internationalizing, country, some referring to specific sectors of the economy, and some referring to MNEs themselves. Whatever the scope of the definition, there are many alternative measures of the extent of internationalization, including production, sales, assets and employment. This chapter emphasizes production as a measure, as far as possible, despite some ambiguities about its meaning.

Five questions are dealt with in this chapter. The first is what, if any, trends there have been in the degree of internationalization of the US economy or of certain sectors of the economy, as measured by the extent of internationalized production relative to home production. The second is how important this internationalization has been, in quantitative terms, for host countries, and particularly for developing host countries. The third question is the extent to which host country sales by affiliates of US firms represent host country production, or the degree of self-sufficiency of affiliates. The fourth question is about the industry composition of US-owned internationalized production, and the fifth is how much, and in what ways, this internationalization has affected developing host countries and possibly the other developing countries that have not been involved in the internationalization process.

## Trends in US production internationalization

A comprehensive measure of the internationalization of the US economy is the size of internationalized production, defined as the gross output of majority-owned US affiliates abroad, relative to output in the US, measured by US

*Table 4.1* Ratio of internationalized US production to US output, 1966–95 (percentages)

| Year | All industries | Manufacturing | Petroleum | Other than petroleum and manufacturing |
|------|------|------|------|------|
| 1966 | 4.89 | n.a. | n.a. | n.a. |
| 1970 | 6.88 | n.a. | n.a. | n.a. |
| 1977 | 8.16 | 15.8 | 129.8 (73.8) | 1.80 |
| 1982 | 7.10 | 15.8 | 58.8 (31.4) | 1.55 |
| 1989 | 6.15 | 17.6 | 77.9 (46.4) | 1.60 |
| 1991 | n.a. | 18.2 | 84.9 | 1.77 |
| 1992 | 5.79 | 17.6 | 99.3 | 1.70 |
| 1993 | 5.62 | 16.4 | 98.0 | 1.67 |
| 1994 | 6.11 | 17.1 | 101.6 | 1.84 |
| 1995 | 6.87 | n.a. | n.a. | n.a. |

*Sources*: Lipsey, Blomström, and Ramstetter (1998); Mataloni and Fahim-Nadar (1996); Yuskavage (1996); US Department of Commerce (1997a); Mataloni and Goldberg (1994); Mataloni (1995), (1997); and US Department of Commerce (1981), (1985), (1992), and (1997b).

GDP. That ratio (as a percentage) for US industry as a whole, rose rapidly in the 1960s and 1970s, reaching a peak in 1977, and then declined, as can be seen in the first column of Table 4.1.

After rising by two thirds from 1966 to 1977, the level of US internationalization declined more than 30 per cent by 1993. It then increased again, and in 1995 was 40 per cent above the 1966 level, but still below the levels of 1977 and 1982. There is little indication of any internationalization juggernaut here.

A similar calculation can be made for the period since 1977 for several broad sectors of the economy, shown in the other columns of Table 4.1. These include petroleum, manufacturing, and all other industries. The last are mainly services but include also agriculture forestry and fishing, as well as mining outside of petroleum.

There was a slight rising trend in the degree of internationalization of US manufacturing through the 1980s, apparently reversed since 1991. By 1994, the ratio was only 7 per cent above the 1977 level, but the increasing trend in internationalization had lasted much longer in manufacturing than in US industry as a whole. Therefore, the period since 1977 has seen little increase in the amount of internationalized US manufacturing production relative to manufacturing production within the United States.

The other large sector for US internationalization is the petroleum industry. The ratios for this industry that we can calculate are biased upward by the fact that the numerators include all petroleum operations abroad, including affiliates not only in extraction and refining, which are in the denominator, but also affiliates in transport, in wholesale and retail trade and in services, which are omitted from the denominator. Even taking these biases into account by assuming that the shares in production of the different segments of the industry are equal to the shares in sales (a questionable assumption in this industry), as shown by the figures in parentheses, it is clear that this is the most internationalized sector of the US economy.

The large drop in internationalization in petroleum between 1977 and 1982 was partly accounted for by foreign government actions. These included Canadian investment policy changes that induced sales of US affiliates to Canadian firms and the acquisition of a large Middle Eastern oil producing affiliate by the host-country government. Since that period the degree of internationalization has been rising, and it remains much greater than in manufacturing. The US petroleum industry's production outside the United States is almost certainly more than half as large as its domestic output.

These two sectors, which account for more than three quarters of US-owned internationalized production, are less than 20 per cent of the US domestic economy, in terms of their production. Most US domestic production is in trade and services, including the distribution and servicing of manufactured products. The implication of these two very different industry distributions is that the US economy outside petroleum and manufacturing is much less internationalized, as seen in Table 4.1.

For three quarters of the US economy, the extent of internationalization of production was negligible, never reaching above 2 per cent, and not showing any distinct trend. In a sense, even these small numbers exaggerate the internationalization of these industries, because much of this trade and service production outside the United States is owned by US manufacturing or petroleum firms and is ancillary to manufacturing or petroleum production.

### Share of developing countries

Only a small part of this internationalized production takes place in developing countries. While almost a third of such production took place in developing countries in 1977, that share was cut to a quarter in 1982 and reached a low point of half the 1977 proportion in 1991. Since then there has been a reversal, with the developing country share increasing, but it remains far below the 1977 level (see Table 4.2). Thus, the main internationalization of US production that has taken place has been a spread into developed economies. That concentration suggests that the main force behind internationalization has not been a search for cheap labour or natural resources, but more likely has reflected the contest for markets, in which high consumer income, despite the accompanying high labour cost, is more of an attraction than low labour costs accompanied by low incomes. The concentration on markets also suggests that barriers to trade, whether natural or artificial, are still important. Most decisions to locate production do not seem to be based mainly on labour costs or other supply side factors.

*Table 4.2* Developing country share of internationalized US production, 1977–95 (percentages)

| Year | All industries | Petroleum | Manufacturing | Others |
|------|----------------|-----------|---------------|--------|
| 1977 | 32.5 | 56.7 | 15.9 | 21.0 |
| 1982 | 25.5 | 32.9 | 20.9 | 20.7 |
| 1989 | 17.5 | 23.6 | 16.5 | 13.3 |
| 1991 | 16.2 | 20.2 | 16.0 | 12.7 |
| 1992 | 18.3 | 21.5 | 18.9 | 13.4 |
| 1993 | 20.8 | 22.9 | 21.4 | 19.6 |
| 1994 | 20.0 | 20.5 | 20.7 | 18.0 |
| 1995 | 20.2 | 23.3 | 21.0 | 18.4 |

*Sources*: Mataloni and Goldberg (1994), Mataloni (1995) and (1997), and Mataloni and Fahim-Nadar (1996).

We might expect that one set of factor prices that would be crucial in location would be prices of natural resources. The geographical breakdown of US internationalized production is available for only one natural resource industry, petroleum, and even that industry has a large share of investment in later stages of production. We cannot separate the stages because the major sub-industry component of the petroleum industry in the US outward foreign direct investment data is 'Integrated Petroleum Refining and Extraction', which makes up 60 per cent of the 'Petroleum' industry.

The early ratios suggest the importance of prices of inputs in determining the location of internationalized production, with more than half located in developing countries despite the concentration of consumption in the developed world. However, by the end of the period, the location of US internationalized petroleum production had become similar to that of other industries, predominantly in developed countries.

Internationalized manufacturing production has been, throughout our period, located mainly in developed countries, presumably close to markets. The share of developing countries has fluctuated between 15 and a little over 20 per cent, with some hint of an upward trend, although the gain since 1991 may well be cyclical.

Outside of petroleum and manufacturing there is no sign of an increasing role for developing countries in US-owned production. There was a sharp drop in the developing country share between 1982 and 1989 and it has stayed much lower than in manufacturing or petroleum since then, although in this case, as in manufacturing, there has been some increase in the developing country share since 1991.

Taking all industries together, the share of developing countries in internationalized US production is considerably lower than their share in world nominal GDP. Developing countries accounted for a little over a third of world GDP from 1970 to 1982 and for a little under 30 per cent in 1989

96

and 1990. In 1993 they accounted for 20 per cent of total internationalized US production and about the same share of internationalized manufacturing production. The difference between developing countries' shares of world output and their shares of US affiliate output emphasizes the influence of high per capita income as an attraction for the location of US affiliate production.

This concentration of US internationalized production in developed countries fits with an earlier finding that for US firms even production for export tended to be concentrated in countries with large domestic markets. That result suggested that economies of scale in producing for large local host country markets influence not only the choice between exporting from the United States and local production by affiliates, but also the choice of affiliate production locations for selling to the world market. The role of scale economies presents a handicap for most developing countries in the competition for such production (Kravis and Lipsey, 1982).

This section can be summarized by the statement that there has been no trend toward increasing internationalization of US firms that has been, or is likely to be in the near future, a force for increased production in developing countries, with the possible exception of the petroleum industry. Within US internationalized production the trend away from production in developing countries seems to have stopped or even been reversed since the early 1990s, particularly in manufacturing and in other industries outside petroleum. The developing country share in manufacturing is about back to its earlier peak. While that is not yet the case in the service industry sector, developing country shares have increased rapidly since the low point around 1990.

## Internationalized US production in developing country output

The fact that internationalized US production is concentrated in developed countries does not mean that it has been unimportant in the output or the growth of developing countries. In particular, it does not mean that it has not been important to individual countries or industries. An overall view of the role of US internationalized production from the point of view of the host countries is presented in Table 4.3, in which data for individual countries are presented where they are available.

The general picture in Table 4.3 is not one of an increasing role for US MNCs in the total production of developing countries since 1977. In developing countries as a group, the US MNC share of output, 3.6 per cent in 1977, fell by more than half through 1991 before seeming to stabilize. More or less the same path can be described for Latin America, Africa and the Middle East, with a particularly large decline in the last case as US petroleum operations were nationalized. The exception was the Asia and Pacific region, where US production shares, lower than in any other region in 1977, did not show

97

*Table 4.3* US MOFAs' share of nominal GDP in developing countries by region and country, 1977–94 (per cent)

| Country/Region | 1977 | 1982 | 1989 | 1991 | 1992 | 1993 | 1994 |
|---|---|---|---|---|---|---|---|
| | | | | *Year* | | | |
| *Developing countries total* | 3.57 | 2.33 | 1.78 | 1.57 | 1.65 | 1.72 | 1.65 |
| *Latin America and Caribbean* | 3.36 | 3.42 | 2.98 | 2.40 | 2.58 | 2.65 | 2.49 |
| Argentina | 2.55 | 3.44 | 2.06 | 1.77 | 1.64 | 1.62 | 1.51 |
| Brazil | 3.68 | 3.98 | 3.70 | 2.84 | 3.53 | 3.48 | 2.63 |
| Chile | 1.21 | 1.92 | 2.42 | 2.69 | 3.06 | 2.68 | 3.29 |
| Colombia | 2.73 | 3.49 | 2.91 | 3.06 | 3.17 | 3.01 | 2.84 |
| Ecuador | 4.61 | 4.15 | 2.77 | 2.78 | 1.34 | 0.92 | 1.33 |
| Peru | 2.84 | 4.50 | 1.52 | 1.16 | 1.20 | 1.12 | 1.31 |
| Venezuela | 3.13 | 3.02 | 1.69 | 2.02 | 2.17 | 2.46 | 2.67 |
| Costa Rica | 3.74 | 6.25 | 3.98 | 3.41 | 6.14 | 5.07 | 6.07 |
| Guatemala | 2.85 | 3.17 | 1.88 | 2.53 | 2.22 | 1.82 | 1.80 |
| Honduras | 8.51 | 8.64 | 8.25 | 9.20 | 8.11 | 6.32 | 9.05 |
| Mexico | 2.35 | 2.05 | 2.37 | 2.61 | 2.59 | 2.40 | 2.62 |
| Panama | 13.91 | 10.01 | 11.42 | 10.21 | 8.98 | 4.36 | 5.12 |
| Jamaica | 11.39 | 13.67 | 11.20 | 9.01 | 8.79 | 6.25 | 6.24 |
| Trinidad and Tobago | – | – | 11.50 | 12.09 | 10.48 | 12.72 | 6.88 |
| *Africa* | 2.99 | 2.10 | 1.20 | 1.29 | 1.28 | 1.42 | 1.28 |
| South Africa | 3.44 | 3.14 | 0.76 | 0.67 | 0.67 | 0.65 | 0.73 |
| Other | 2.90 | 1.84 | 1.33 | 1.52 | 1.51 | 1.73 | 1.51 |
| *Middle East* | 10.93 | 2.29 | 1.56 | 0.79 | 1.01 | 0.85 | 0.81 |
| Israel | 1.49 | 1.07 | 0.77 | 1.00 | 1.01 | 1.11 | 1.19 |
| Other | 11.69 | 2.38 | 1.70 | 0.75 | 1.01 | 0.79 | 0.71 |
| *Asia and Pacific* | 1.30 | 1.46 | 1.16 | 1.23 | 1.22 | 1.27 | 1.25 |
| China | 0.00 | 0.00 | 0.00 | 0.06 | 0.09 | 0.08 | 0.13 |
| Hong Kong | 3.47 | 3.02 | 4.36 | 3.71 | 3.46 | 3.62 | 3.72 |
| India | 0.19 | 0.12 | 0.06 | 0.05 | 0.05 | 0.07 | 0.08 |
| Indonesia | 9.63 | 6.67 | 4.23 | 4.31 | 3.98 | 3.08 | 2.66 |
| Korea, Republic of | 0.21 | 0.29 | 0.33 | 0.35 | 0.35 | 0.34 | 0.39 |
| Malaysia | 2.43 | 6.31 | 4.62 | 4.28 | 5.37 | 4.94 | 5.07 |
| Philippines | 2.79 | 2.89 | 2.36 | 2.62 | 2.67 | 2.89 | 2.82 |
| Singapore | 6.08 | 7.26 | 8.07 | 7.88 | 6.66 | 8.33 | 8.34 |
| Taiwan | 1.19 | 1.29 | 1.30 | 1.33 | 1.26 | 1.11 | 1.17 |
| Thailand | 1.28 | 1.80 | 2.51 | 2.23 | 1.91 | 1.92 | 1.85 |

*Sources*: Appendix Tables 4.A1 and 4.A2.

*Note*: MOFA is Majority Owned Foreign Affiliate.

any particular trend. US affiliates' shares of host country production increased in five developing countries and decreased in twelve outside Asia, but they increased in seven and decreased in only three Asian countries, even though total production rose more rapidly in the Asian countries than in the other regions.

In only nine out of the twenty-eight countries (including two country groups) for which we have data in Table 4.3 did the share of US internationalized production ever reach as high as 5 per cent of total output. Three of these were oil producers, Trinidad and Tobago, the Middle East other than Israel, and Indonesia; and four were small countries in the Western Hemisphere. The other two were Malaysia and Singapore, the former with US-owned production 4 to 6 per cent of total output and the latter with 6 to 9 per cent. Considering that both countries have been hosts to substantial Japanese investment, and that the Japanese shares have almost certainly been increasing, these are probably two cases of an important role for internationalized production in host country growth.

Apart from the recent interest in infrastructure investment, the strongest impact of internationalized production might be expected in the manufacturing sector. A summary of US firms' foreign operations in that sector is presented in Table 4.4. The problem mentioned above with respect to comparisons between affiliate production and domestic production in the US exists here too. The numerators exclude petroleum refining while the denominators include it in manufacturing.

US affiliates accounted for between 2.5 and 5 per cent of developing country manufacturing output, with a peak in 1982 and a sharp decline by 1994. The US firm share was twice as high in Latin America with the same time pattern, but lower elsewhere. Some of the highest US ownership ratios were in small Latin American or Caribbean countries with very little manufacturing, but more significant high ratios in that region were in Brazil and in Venezuela, often, but not always, above 10 per cent. In Asia, the fastest growing manufacturing region, US MNE shares were high in Hong Kong and Malaysia and highest in Singapore, in the latter case reaching over 20 per cent. All of these countries showed sings of an upward trend in the US affiliate share of manufacturing output.

Some perspective on these shares of US-owned production in host country output is given by the totals for all foreign-owned production in some of the countries, especially Asian countries where Japan, rather than the United States, is the leading investor. Most of the data refer to manufacturing, and give a picture of the wide range of experience with internationalized production. Malaysia and Singapore are extreme examples on one side. Foreign-owned production in Malaysia has accounted for over 40 per cent of manufacturing output in recent years, after a twenty-year decline from more than half during the 1970s. In Singapore, the foreign share was over 60 per cent during the 1960s and has since grown to more than 70 per cent. In Taiwan, the foreign share has usually been over 15 per cent and, in most recent years, over 20 per cent. Foreign shares have mostly been in the 10 to 15 per cent range in Hong Kong (with a rising trend), Indonesia (with a falling trend from the 20 to 30 per cent shares of the 1970s and early 1980s), Korea and Thailand. The smallest foreign shares in manufacturing in countries for which

*Table 4.4* US MOFAs' share of nominal manufacturing gross output in developing countries, by region and country, 1977–94 (per cent)

| Country/Regions | Year | | | | | | |
|---|---|---|---|---|---|---|---|
| | *1977* | *1982* | *1989* | *1991* | *1992* | *1993* | *1994* |
| *Developing countries total* | 4.40 | 4.88 | 3.18 | – | – | – | 2.88 |
| *Latin America and Caribbean* | 8.34 | 9.70 | 8.92 | – | – | – | 6.74 |
| Argentina | 4.55 | 7.03 | 4.10 | 4.00 | 4.56 | – | 3.75 |
| Brazil | 11.35 | 10.94 | 11.72 | 10.98 | – | – | 9.47 |
| Chile | 2.13 | 2.66 | 5.56 | – | – | – | 5.42 |
| Colombia | 6.89 | 8.71 | 7.87 | 7.64 | 7.69 | – | 4.86 |
| Ecuador | 2.92 | 3.26 | 1.79 | 1.35 | 1.25 | 1.35 | 2.22 |
| Peru | 2.49 | – | – | – | – | 0.90 | 1.03 |
| Venezuela | 11.15 | 11.11 | 7.24 | 8.64 | 10.77 | 12.49 | 11.08 |
| Costa Rica | 9.77 | 12.83 | 9.31 | 12.02 | 10.51 | 10.35 | 11.74 |
| Guatemala | 9.19 | 7.36 | 7.92 | 6.99 | – | – | 7.05 |
| Honduras | 14.90 | 21.35 | 22.45 | 27.35 | 25.70 | 15.03 | 30.29 |
| Mexico | 8.40 | 7.99 | 8.18 | 10.22 | 11.04 | 10.68 | 10.93 |
| Panama | 11.11 | 11.42 | 51.46 | 40.23 | 40.06 | 11.97 | 11.04 |
| Jamaica | 3.00 | 22.46 | 10.67 | 14.78 | 9.68 | 7.65 | 13.32 |
| Trinidad and Tobago | – | 4.80 | 3.86 | 2.27 | 2.42 | 2.59 | 0.90 |
| *Africa* | 3.83 | 4.06 | 2.14 | – | – | – | 2.35 |
| South Africa | 6.86 | 5.99 | 2.13 | 1.83 | 1.89 | 1.66 | 2.12 |
| Other | 1.97 | 2.06 | 2.15 | – | – | – | 2.66 |
| *Middle East* | 0.79 | 0.80 | 0.71 | – | – | – | 2.14 |
| Israel | 2.81 | 3.20 | 3.10 | – | – | – | 6.37 |
| Other | 0.19 | 0.32 | 0.02 | – | – | – | 0.22 |
| *Asia and Pacific* | 1.22 | 1.33 | 1.03 | – | – | – | 1.26 |
| China | 0.00 | 0.01 | 0.03 | 0.06 | 0.08 | 0.14 | 0.25 |
| Hong Kong | 5.70 | 3.92 | 6.09 | 6.49 | 7.14 | 7.60 | 9.50 |
| India | 1.19 | 0.72 | 0.35 | 0.33 | 0.32 | 0.46 | 0.46 |
| Indonesia | 2.08 | 1.29 | 0.58 | 0.47 | 0.52 | 0.49 | 0.93 |
| Korea, Republic of | 0.58 | 0.56 | 0.67 | 0.71 | 0.74 | 0.70 | 0.70 |
| Malaysia | 4.83 | 7.63 | 5.61 | 7.08 | 9.42 | 8.28 | 10.78 |
| Philippines | 5.56 | 4.80 | 5.90 | 5.98 | 6.62 | 7.18 | 6.50 |
| Singapore | 12.86 | 14.96 | 16.98 | 15.55 | 16.24 | 20.47 | 20.06 |
| Taiwan | 3.00 | 3.07 | 2.97 | 2.62 | 2.59 | 2.08 | 2.14 |
| Thailand | 1.45 | 1.23 | 2.46 | 1.69 | 1.44 | 1.49 | 1.91 |

*Sources*: Appendix Tables 4.A3 and 4.A4.

we have data, have been in India, where they have mostly been 6 to 7 per cent. The most rapid growth seems to have taken place in Guangdong Province of China, where the foreign share in manufacturing must have been under 2 per cent in 1980 but had passed 8 per cent in 1990 and reached a third in 1993,

100

accounting for more than half of the very rapid growth in manufacturing output (Lipsey, Blomström and Ramstetter, 1998).

## The self-sufficiency of US affiliates

One way in which US MNEs' policies could affect host countries would be a change in the degree of self-sufficiency of affiliates, as measured by the proportion of their sales supplied from their own production in the host country. Information is available for production taking place within the host country by the affiliates themselves, although not for any host country products produced by others and purchased by the affiliates.

For all industries, the share of affiliate sales in developing countries accounted for by the affiliates' own production was about a third in 1977 and declined to a little over a quarter in 1994 and 1995 (Table 4.5). The sharpest decline in that ratio took place in Asia, while Latin America and the Caribbean showed a slight trend in the opposite direction, moving a little towards greater self-sufficiency on the part of US affiliates in the area as a whole. However, in most individual Latin American and Caribbean countries, there has been a decline in US affiliate self-sufficiency since 1989, just as there was in Asia.

Within manufacturing, the ratios of production to sales among US affiliates were relatively high: almost always close to or above one third (Table 4.6). Latin American manufacturing affiliates have been especially self-sufficient, the ratios at or above 40 per cent, with no trend, although there was a dip in 1994. The explanation may be that Latin American trade policies have been more autarkic than those of other regions. Asian manufacturing affiliates, in contrast, have produced smaller proportions of what they have sold, and the proportion has been declining on average, to only about 60 per cent of the Latin American level in the most recent year. Asian manufacturing affiliates are probably more involved in worldwide integrated production arrangements in which they perform specific parts of a chain of production, as in the semiconductor industry. The effect of a restrictive trade regime probably accounts for the high ratios of production to sales in India, closer to those of Latin America than to those of other Asian countries, at least until 1994.

If most sales by US affiliates were not coming from their own production in the host countries, where were they coming from? One possibility is that the affiliates were largely, and increasingly, reselling parent output. Overall there was some trend towards supplying host country sales by importing from parents; the ratio of imports from parents to sales more than doubled, reaching over 11 per cent by 1994 (Table 4.7). Simple reselling of parent output was clearly not the main activity of most affiliates, although in a few countries and periods, resales, including the affiliates' markup, of

*Table 4.5* US MOFAs' output as percentage of sales in developing countries, 1977–95 (per cent)

| Country/Region | Year | | | | | | | |
|---|---|---|---|---|---|---|---|---|
| | 1977 | 1982 | 1989 | 1991 | 1992 | 1993 | 1994 | 1995 |
| *Developing countries total* | 32.94 | 29.49 | 31.37 | 27.11 | 27.61 | 29.35 | 26.95 | 26.65 |
| *Latin America and Caribbean* | 27.55 | 26.90 | 30.51 | 27.64 | 29.13 | 32.71 | 30.91 | 30.71 |
| Argentina | 40.08 | 56.86 | 38.87 | 59.46 | 49.13 | 50.66 | 36.77 | 37.45 |
| Brazil | 39.00 | 43.00 | 54.33 | 42.66 | 51.97 | 55.45 | 50.63 | 46.46 |
| Chile | 30.39 | 35.92 | 34.38 | 37.40 | 37.74 | 31.94 | 34.78 | 32.22 |
| Colombia | 28.46 | 30.92 | 29.53 | 29.55 | 30.93 | 28.82 | 28.15 | 28.71 |
| Ecuador | 59.73 | 64.26 | 47.06 | 45.42 | 27.66 | 22.94 | 27.67 | 16.13 |
| Peru | 46.60 | 61.05 | 35.38 | 24.13 | 28.23 | 32.88 | 40.44 | 43.39 |
| Venezuela | 29.07 | 33.07 | 27.49 | 30.57 | 30.48 | 31.37 | 29.00 | 32.11 |
| Costa Rica | – | – | 28.77 | 28.24 | 27.24 | 24.57 | 28.24 | 29.47 |
| Guatemala | – | – | 23.51 | 32.47 | 28.71 | 28.14 | 27.29 | 24.61 |
| Honduras | – | – | 24.30 | 32.24 | 27.84 | 21.74 | 27.63 | 30.08 |
| Mexico | 34.98 | 31.60 | 29.71 | 30.54 | 28.75 | 26.99 | 24.98 | 21.55 |
| Panama | 21.57 | 14.22 | 29.04 | 31.98 | 27.99 | 16.81 | 19.09 | 22.39 |
| Jamaica | – | – | 39.98 | 39.99 | 28.84 | 25.29 | 23.66 | 30.53 |
| Trinidad and Tobago | – | – | 76.23 | 76.07 | – | – | 43.42 | 59.89 |
| *Africa* | 35.24 | 32.74 | 39.72 | 39.38 | 38.54 | 44.22 | 36.40 | 37.62 |
| South Africa | 28.55 | 29.72 | 26.42 | 24.42 | 25.76 | 25.19 | 25.01 | 21.63 |
| Other | 37.38 | 34.24 | 43.67 | 43.80 | 42.15 | 49.65 | 40.08 | 44.17 |
| *Middle East* | 35.38 | 48.58 | 60.98 | 34.91 | 39.39 | 38.60 | 38.05 | 38.63 |
| Israel | 39.61 | 47.54 | 34.45 | 39.85 | 40.89 | 39.34 | 39.73 | 37.58 |
| Other | 35.34 | 48.62 | 64.94 | 33.72 | 39.04 | 38.35 | 37.37 | 39.05 |
| *Asia and Pacific* | 39.21 | 26.92 | 27.35 | 23.90 | 23.23 | 23.23 | 21.34 | 21.41 |
| China | – | – | 3.11 | 21.25 | 26.63 | 23.91 | 21.02 | 18.81 |
| Hong Kong | 10.99 | 12.76 | 17.83 | 16.08 | 16.03 | 16.99 | 16.48 | 17.52 |
| India | 40.94 | 37.06 | 48.61 | 40.07 | 36.67 | 39.11 | 23.60 | 27.24 |
| Indonesia | 85.24 | 50.36 | 65.34 | 63.76 | 64.05 | 62.74 | 56.50 | 58.39 |
| Korea, Republic of | 36.07 | 36.26 | 29.48 | 29.51 | 27.42 | 27.17 | 26.14 | 25.63 |
| Malaysia | 29.01 | 39.15 | 32.28 | 27.19 | 35.48 | 32.83 | 30.91 | 30.38 |
| Philippines | 26.62 | 29.87 | 34.63 | 32.88 | 34.62 | 35.98 | 34.60 | 32.59 |
| Singapore | 18.93 | 7.86 | 15.58 | 11.77 | 9.68 | 12.44 | 12.27 | 12.74 |
| Taiwan | 25.90 | 32.99 | 28.61 | 30.35 | 29.60 | 24.49 | 20.53 | 19.94 |
| Thailand | 27.73 | 25.37 | 33.27 | 29.51 | 28.42 | 29.43 | 27.46 | 27.30 |

*Sources*: Appendix Tables 4.A2 and 4.A5.

products partly or wholly produced by parents, must have been a major activity. The importance of parent sales was originally lower in Latin America than in Asia, possibly because there were more restrictions on importation. By the 1990s, however, it was higher in Latin America, mainly the result of a very large increase in Mexico to a level twice as high as in the next highest country.

*Table 4.6* US manufacturing MOFAs' output of manufacturing as percentage of sales in developing countries, 1977–94 (per cent)

| | *Year* | | | | | | |
|---|---|---|---|---|---|---|---|
| *Country/Region* | *1977* | *1982* | *1989* | *1991* | *1992* | *1993* | *1994* |
| *Developing countries total* | 37.04 | 40.29 | 37.86 | 32.90 | 34.23 | 32.74 | 30.31 |
| *Latin America and Caribbean* | 39.36 | 44.38 | 44.91 | 39.55 | 40.84 | 39.94 | 35.98 |
| Argentina | 38.32 | 57.68 | 35.81 | 53.59 | 45.45 | 52.23 | 32.25 |
| Brazil | 46.08 | 56.18 | 58.23 | 47.87 | 57.92 | 55.19 | 51.77 |
| Chile | 31.16 | 31.85 | 37.53 | 39.49 | 37.39 | 32.19 | 36.72 |
| Colombia | 32.26 | 39.00 | 42.10 | 38.22 | 35.56 | 32.27 | 32.40 |
| Ecuador | 30.70 | 39.78 | 19.79 | 14.80 | 14.40 | 17.14 | 26.58 |
| Peru | 23.87 | 33.75 | 32.26 | 29.17 | 27.95 | 24.05 | 25.55 |
| Venezuela | 26.67 | 28.90 | 25.66 | 27.58 | 28.20 | 29.53 | 27.73 |
| Costa Rica | – | – | 32.04 | 30.82 | 26.75 | 25.25 | 29.13 |
| Guatemala | – | – | 35.03 | 35.03 | 34.05 | 35.41 | 31.25 |
| Honduras | – | – | 25.06 | 38.89 | 35.31 | 19.15 | 38.12 |
| Mexico | 35.18 | 30.50 | 28.94 | 30.30 | 28.83 | 28.32 | 26.14 |
| Panama | 36.62 | 26.01 | 98.91 | 92.39 | 45.48 | 24.90 | 36.55 |
| Jamaica | – | – | 34.98 | 74.10 | 36.81 | 29.03 | 43.38 |
| Trinidad and Tobago | – | 36.49 | 26.23 | 24.44 | 24.49 | 23.40 | −14.29 |
| *Africa* | 31.07 | 31.08 | 34.49 | 29.77 | 31.44 | 29.41 | 30.18 |
| South Africa | 28.14 | 29.75 | 32.50 | 30.78 | 32.45 | 28.54 | 29.23 |
| Other | 39.94 | 35.91 | 36.74 | 28.76 | 30.46 | 30.21 | 31.26 |
| *Middle East* | 34.92 | 37.78 | 36.25 | 39.88 | 41.55 | 39.35 | 39.69 |
| Israel | 40.19 | 41.20 | 39.38 | 41.17 | 42.84 | 39.11 | 42.32 |
| Other | 22.09 | 32.47 | 7.55 | 27.47 | 27.66 | 43.42 | 21.83 |
| *Asia and Pacific* | 29.17 | 28.18 | 24.64 | 22.21 | 23.05 | 21.47 | 22.00 |
| China | – | – | 29.75 | 21.94 | 23.06 | 30.40 | 25.82 |
| Hong Kong | 26.71 | 21.67 | 21.20 | 19.46 | 19.80 | 18.68 | 19.57 |
| India | 43.90 | 41.14 | 59.41 | 48.87 | 44.36 | 51.14 | 29.32 |
| Indonesia | 40.46 | 30.17 | 29.33 | 25.62 | 28.38 | 27.27 | 22.58 |
| Korea, Republic of | 31.55 | 28.74 | 30.50 | 27.95 | 26.65 | 27.29 | 25.48 |
| Malaysia | 27.42 | 23.05 | 17.79 | 21.07 | 25.02 | 21.03 | 24.43 |
| Philippines | 27.52 | 26.64 | 37.56 | 33.19 | 33.44 | 33.06 | 31.93 |
| Singapore | 23.81 | 30.37 | 19.17 | 16.84 | 17.19 | 18.36 | 18.37 |
| Taiwan | 28.64 | 34.36 | 31.38 | 31.98 | 33.14 | 23.90 | 25.79 |
| Thailand | 24.79 | 18.43 | 22.33 | 16.75 | 16.01 | 17.14 | 20.38 |

*Sources*: Appendix Tables 4.A4 and 4.A6.

Despite the rise for Latin America as a whole, there was a predominance of decreases in dependence on parents in the individual countries.

Asian affiliates' dependence on imports from parents increased much less, and decreases were predominant among individual countries. The largest decreases were in countries with the highest initial levels.

*Table 4.7* US MOFAs' imports from parents as percentage of sales in developing countries, 1977–94 (per cent)

| Country/Region | Year | | | | | | |
|---|---|---|---|---|---|---|---|
| | *1977* | *1982* | *1989* | *1991* | *1992* | *1993* | *1994* |
| *Developing countries total* | 4.29 | 6.51 | 9.59 | 9.87 | 9.80 | 10.87 | 11.54 |
| *Latin America and Caribbean* | 6.36 | 6.24 | 9.79 | 12.42 | 12.22 | 12.84 | 14.00 |
| Argentina | 5.15 | 6.21 | 5.99 | 6.29 | 4.82 | 6.38 | 5.67 |
| Brazil | 4.35 | 2.71 | 4.06 | 3.88 | 3.96 | 4.45 | 4.90 |
| Chile | 9.94 | 4.83 | – | 6.46 | 4.47 | 5.44 | 4.94 |
| Colombia | 9.68 | 5.43 | 6.44 | 5.57 | 6.03 | 6.10 | 8.00 |
| Ecuador | 7.78 | 10.46 | 13.84 | 10.97 | 8.51 | 7.01 | 5.66 |
| Peru | 9.57 | 14.50 | 2.50 | 2.77 | 2.89 | 4.07 | 5.09 |
| Venezuela | 22.11 | 18.85 | 12.18 | 19.87 | 20.26 | 23.37 | 16.07 |
| Costa Rica | – | – | 10.65 | 6.03 | 6.20 | 6.03 | 6.40 |
| Guatemala | – | – | 7.59 | 16.10 | 15.22 | 4.37 | 10.82 |
| Honduras | – | – | 11.69 | 12.85 | 11.65 | 8.62 | 8.70 |
| Mexico | 13.87 | 20.66 | 37.05 | 35.80 | 33.50 | 32.93 | 33.89 |
| Panama | 5.90 | 7.85 | 11.18 | 15.79 | 8.76 | 9.29 | 11.26 |
| Jamaica | – | – | 11.60 | 15.52 | 7.82 | 6.85 | 14.43 |
| Trinidad and Tobago | 2.77 | 2.98 | 1.69 | 0.71 | – | – | 2.37 |
| *Africa* | 3.41 | 4.23 | – | 2.49 | 2.17 | 2.33 | 2.89 |
| South Africa | 7.02 | 6.68 | 6.63 | 6.01 | 5.36 | 5.09 | 6.34 |
| Other | 2.25 | 3.01 | – | 1.45 | 1.26 | 1.54 | 1.77 |
| *Middle East* | 1.49 | 3.78 | 3.59 | 2.32 | 2.01 | 2.08 | 2.45 |
| Israel | 4.93 | 9.34 | 12.67 | 2.14 | 2.09 | 1.57 | 2.89 |
| Other | 1.46 | 3.58 | 2.24 | 2.36 | 2.00 | 2.25 | 2.27 |
| *Asia and Pacific* | 8.16 | 9.10 | 11.03 | 8.72 | 8.79 | 10.43 | 10.57 |
| China | – | – | 13.62 | 10.78 | 10.98 | 17.66 | 6.11 |
| Hong Kong | 6.57 | 12.45 | 12.98 | 9.54 | 12.63 | 16.15 | 16.72 |
| India | 2.14 | 1.94 | 6.81 | 6.51 | 5.45 | 4.22 | 2.24 |
| Indonesia | 2.32 | 3.09 | 1.55 | 1.13 | 2.05 | 3.60 | 4.05 |
| Korea, Republic of | 45.21 | 28.48 | 17.13 | 17.37 | 16.04 | 20.41 | 16.96 |
| Malaysia | 19.25 | 22.44 | 12.33 | 8.91 | 8.47 | 9.56 | 9.17 |
| Philippines | 6.40 | 8.23 | 8.23 | 4.04 | 3.09 | 3.27 | 4.87 |
| Singapore | 16.23 | 6.77 | 13.69 | 8.44 | 7.30 | 8.21 | 9.82 |
| Taiwan | 16.53 | 20.62 | 10.22 | 11.43 | 11.64 | 11.82 | 8.85 |
| Thailand | 8.19 | 11.43 | 8.12 | 10.35 | 8.78 | 8.58 | 7.74 |

*Sources*: Appendix Tables 4.A5 and 4.A7.

Manufacturing affiliates are considerably more import-dependent than those in other industries, and that dependence has, in the aggregate, increased a little since 1977 (Table 4.8). There was an upward trend in the import dependence of manufacturing affiliates in Latin America, reflecting mainly

*Table 4.8* US manufacturing MOFAs' imports from parents as percentage of sales in developing countries, 1977–94 (per cent)

| Country/Region | Year | | | | | | |
|---|---|---|---|---|---|---|---|
| | *1977* | *1982* | *1989* | *1991* | *1992* | *1993* | *1994* |
| *Developing countries total* | 11.97 | 13.99 | 15.22 | 16.32 | 15.78 | 15.15 | 14.98 |
| *Latin America and Caribbean* | 10.92 | 11.08 | 15.78 | 19.28 | 19.00 | 18.25 | 19.08 |
| Argentina | 6.33 | 7.76 | 6.99 | 5.24 | 3.61 | 4.57 | 3.36 |
| Brazil | 5.51 | 3.45 | 4.87 | 4.74 | 4.98 | 4.29 | 5.04 |
| Chile | 10.05 | 5.22 | – | 5.22 | 4.26 | 4.04 | 5.03 |
| Colombia | 13.21 | 6.55 | 7.84 | 6.92 | 7.78 | 6.81 | 8.72 |
| Ecuador | – | – | – | – | 10.70 | 7.76 | 6.64 |
| Peru | 7.42 | – | – | 9.26 | 8.27 | 8.23 | 6.57 |
| Venezuela | 26.82 | 17.74 | 10.13 | 19.47 | 20.08 | 25.74 | 16.72 |
| Costa Rica | – | – | 9.71 | 6.85 | 5.54 | 5.25 | 15.26 |
| Guatemala | – | – | 5.58 | 6.60 | 6.47 | 5.84 | 4.28 |
| Honduras | – | – | 11.46 | 4.25 | 5.39 | 4.98 | 2.61 |
| Mexico | 15.49 | 22.35 | 39.78 | 38.17 | 36.28 | 35.70 | 38.15 |
| Panama | 23.94 | 20.81 | – | – | 4.52 | 9.73 | 14.66 |
| Jamaica | – | – | – | 13.67 | 18.40 | 14.52 | 11.87 |
| Trinidad and Tobago | – | 22.97 | – | – | 8.16 | 6.38 | – |
| *Africa* | 9.10 | 8.94 | 8.79 | 5.30 | 4.75 | 4.28 | 6.46[a] |
| South Africa | 7.22 | 8.45 | – | 8.27 | 0.26 | 6.70 | 5.97 |
| Other | 14.82 | 10.75 | – | 2.34 | 9.07 | 2.04 | – |
| *Middle East* | 14.58 | 10.91 | 2.97 | 2.91 | 2.09 | 1.92 | |
| Israel | – | 10.96 | – | – | – | 0.70 | 2.19 |
| Other | – | 10.82 | – | – | – | 22.37 | – |
| *Asia and Pacific* | 18.24 | 27.90 | 15.05 | 12.91 | 11.69 | 11.44 | 10.24 |
| China | – | – | 9.09 | 4.27 | 8.51 | – | 5.61 |
| Hong Kong | 16.91 | 19.30 | 11.74 | 14.53 | 12.50 | 11.54 | 9.55 |
| India | 1.71 | 2.17 | 1.48 | 2.63 | 2.18 | 1.71 | 1.76 |
| Indonesia | 11.07 | 21.90 | – | 8.54 | 5.09 | 4.31 | 4.00 |
| Korea, Republic of | 47.59 | 35.51 | 15.74 | 15.93 | 17.15 | 18.42 | 15.38 |
| Malaysia | 44.94 | 54.88 | 21.26 | 13.82 | 12.39 | 13.76 | 11.36 |
| Philippines | 6.44 | 14.54 | 11.66 | 5.26 | 3.23 | 3.50 | 5.80 |
| Singapore | 24.72 | 29.89 | – | 13.04 | 12.42 | 11.14 | 10.25 |
| Taiwan | 18.16 | 23.26 | 10.10 | 9.62 | 9.75 | 10.49 | 10.86 |
| Thailand | 20.94 | 44.34 | 14.73 | 21.47 | 16.26 | 16.27 | 15.19 |

*Sources*: Appendix Tables 4.A6 and 4.A8.

*Notes*: *a* includes Middle East.

the rise in Mexico, but downward trends in the other regions and in almost all the individual countries for which data are available. In Mexico, the only country in which the recent import dependence levels are much higher than the early ones, the major changes took place between 1977 and 1982 and especially between 1982 and 1989, with only minor fluctuations after that.

There was a downward trend in Asia, but again, most individual country changes were reductions in dependence on imports from parents. In particular, in countries where import dependence was extremely high, such as Korea and Malaysia, manufacturing affiliates had greatly reduced that dependence by the 1990s. That reduction in dependence on parents did not mean a move towards self-sufficiency, as was seen in Table 4.6, but rather a shift in sources of supply.

Affiliate production and, for the most part, imports from parents, represent production within the multinational firm. The sum of the two changed little over the period, as seen in Table 4.9.

Production within the affiliates themselves and imports from their parents usually supplied a little under 40 per cent of affiliate sales in developing countries, leaving over 60 per cent to be supplied by local purchases or by imports from non-parent sources. We have no information about the division between local purchases and imports. However, we can make a rough approximation to inputs produced by other related affiliates. For imported inputs we use exports by affiliates in all countries to affiliates in developing countries and for locally produced related firm inputs we use local sales to other affiliates by affiliates in developing countries.

For affiliates in all industries combined, the purchases from other local affiliates do not seem to have been a growing source of affiliate sales. For three of these years there are also data on sales by affiliates to other affiliates located in developing countries, another source of inputs internal to the MNCs. These also amounted to a rather small percentage of developing country affiliate sales. Thus, for all industries combined, inputs internal to MNCs added up to about half of sales of US affiliates in developing countries. In

Table 4.9 Proportion of affiliate sales in developing countries accounted for by local production and related imports, 1977–94

| Year | All Industries | | | | Manufacturing | | |
|------|---------------------------------------------------|---------------------------------|-----------------------------------|-------|---------------------------------------------------|-----------------------------------|-------|
| | Affiliate production and imports from parents | Affiliated local purchases | Imports from other affiliates | Total | Affiliate production and imports from parents | Imports from other affiliates | Total |
| 1977 | 38.1 | 7.9 | 7.8 | 53.9 | 49.0 | 7.8 | 56.8 |
| 1982 | 37.2 | 3.3 | 4.8 | 45.3 | 54.3 | 6.9 | 61.2 |
| 1989 | 41.0 | 5.4 | 8.2 | 54.6 | 53.1 | 8.9 | 62.0 |
| 1994 | 38.5 | 4.1 | n.a. | n.a. | 45.3 | n.a. | n.a. |

Sources: Tables 4.5, 4.6, 4.7, and 4.8 and US Department of Commerce (1981), (1985), (1992) and (1997a).

other words, about half of the sales of US MNC affiliates were either from their own output in the host countries or were produced elsewhere by their parents or by other affiliates of the same parents. That proportion was substantially lower in 1982 than in the preceding or following years, but no trend is evident.

Manufacturing affiliates, as noted above, are more self-sufficient than affiliates in general, and less dependent on inputs from outside their own production and their parents' exports to them. If we add to those two types of inputs their imports from other affiliates of the same parents (we do not have data on purchases from sibling affiliates in the same country), we find that manufacturing affiliates in developing countries produced themselves, or purchased from within their MNC networks, about 60 per cent of what they sold, leaving 40 per cent for purchases from other host country firms or from non-affiliated firms outside their host countries. If these affiliates had the same propensity to purchase from other related affiliates in their own host countries as affiliates in other industries (see above), the shares from unrelated suppliers might be as low as a third, although no clear trend emerges.

## The industry distribution of affiliate manufacturing production

Host countries differ not only in the importance of manufacturing production by US affiliates, but also in the kind of manufacturing they attract. The industry distributions of such production in the countries for which the data are available are given in Table 4.10. The distributions do not differ greatly between developed and developing countries except that, within machinery, industrial machinery is more important in developed country affiliates and electrical machinery in developing country affiliates. There is a much larger contrast within developing country affiliates between those in Latin America and those in Asia. The Latin American production is much more concentrated in foods, chemicals and transport equipment (62 per cent as against 24 per cent) and that of the Asian countries in electrical equipment and non-electrical machinery (60 per cent as against 11 per cent). The Asian countries listed, with the exception of India and Indonesia, all had at least 20 per cent of US affiliate production in the electrical machinery industry, while the highest share for that industry in Latin America was 15 per cent in Mexico, and all the other shares that can be calculated were 3 per cent or less.

Another apparent pattern is that high manufacturing production shares in chemicals seem to be associated with petroleum or other mining production. Thus, Colombia, Ecuador, Peru, South Africa, Indonesia and the Philippines all had 30 per cent or more of US MOFA production in chemicals. However, the relationship is not a close one. Some countries with substantial mining

*Table 4.10* Industry distribution of manufacturing gross product of US affiliates, 1994

| Country / Region | Total Mfg | Food and kindred products | Chemicals and allied products | Primary and fabricated metals | Industrial machinery and equipt | Electronic and other electrical equipt | Transport equipt | Other Mfg |
|---|---|---|---|---|---|---|---|---|
| *World* | 100.0 | 12.1 | 20.0 | 3.9 | 13.4 | 9.7 | 17.5 | 23.5 |
| *Developed countries* | 100.0 | 11.6 | 19.9 | 4.0 | 14.2 | 8.2 | 17.8 | 24.3 |
| *Developing countries* | 100.0 | 13.9 | 20.4 | 3.7 | 10.3 | 15.4 | 16.1 | 20.1 |
| Latin America and Caribbean | 100.0 | 17.1 | 23.0 | 4.0 | 4.1 | 6.6 | 22.1 | 23.0 |
| Argentina | 100.0 | 42.9 | 28.1 | 3.0 | 2.1 | 2.0 | 1.3 | 20.6 |
| Brazil | 100.0 | 9.9 | 20.9 | 4.1 | 6.1 | 2.9 | 29.4 | 26.7 |
| Chile | 100.0 | 11.9 | 21.8 | 32.3 | b | 1.2 | 12.6 | 20.2 |
| Colombia | 100.0 | 20.9 | 35.2 | 3.1 | −0.1 | b | b | 23.7 |
| Ecuador | 100.0 | 46.3 | 30.0 | 5.0 | 0.0 | b | 0.0 | 17.5 |
| Peru | 100.0 | 23.8 | 44.8 | 1.9 | 0.0 | 2.9 | 0.0 | 27.6 |
| Venezuela | 100.0 | 21.4 | 21.4 | 4.1 | 1.9 | b | b | 30.1 |
| Costa Rica | 100.0 | 31.0 | 21.9 | 7.5 | 0.0 | b | 0.0 | b |
| Guatemala | 100.0 | 28.4 | 22.1 | 6.3 | 0.0 | 0.0 | 0.0 | 43.2 |
| Honduras | 100.0 | b | 2.1 | 1.4 | 0.0 | 0.0 | 0.0 | b |
| Mexico | 100.0 | 17.8 | 23.7 | 2.2 | 4.1 | 16.1 | 21.0 | 15.2 |
| Panama | 100.0 | 18.6 | 28.6 | 5.7 | 0.0 | 0.0 | 0.0 | 48.6 |
| Jamaica | 100.0 | 0.0 | 48.4 | 0.0 | 0.0 | 0.0 | 0.0 | 51.6 |
| *Africa* | 100.0 | 28.4 | 34.6 | 9.9 | 3.7 | 1.1 | 2.0 | 20.3 |
| South Africa | 100.0 | 16.1 | 45.8 | 6.0 | 6.6 | 0.9 | 3.8 | 21.2 |
| Other | 100.0 | 41.4 | 22.8 | 13.9 | 0.6 | 1.4 | 0.0 | 19.3 |
| *Middle East* | 100.0 | 12.7 | 6.4 | 1.1 | 0.8 | 67.3 | 0.3 | 11.5 |
| Israel | 100.0 | b | 4.4 | 0.5 | 0.2 | 71.6 | 0.0 | b |
| Other | 100.0 | b | 32.0 | 10.0 | 10.0 | 10.0 | 0.0 | b |
| *Asia and Pacific* | 100.0 | 5.2 | 14.0 | 2.5 | 25.9 | 33.9 | 4.5 | 14.0 |
| China | 100.0 | 5.9 | 22.1 | 6.8 | 17.4 | 35.2 | 0.0 | 12.7 |
| Hong Kong | 100.0 | 1.4 | 6.6 | 5.8 | 9.5 | 44.5 | 1.6 | 30.7 |
| India | 100.0 | b | 37.8 | b | 20.3 | −0.9 | 0.0 | 36.9 |
| Indonesia | 100.0 | 4.1 | 65.1 | 2.1 | b | 4.9 | 2.6 | 21.2ª |
| Korea, Republic of | 100.0 | 11.2 | 17.2 | 1.0 | 6.8 | 30.3 | 2.8 | 30.7 |
| Malaysia | 100.0 | 1.2 | 4.9 | b | 13.4 | 63.7 | 0.0 | 32.5ᶜ |
| Philippines | 100.0 | 24.8 | 36.4 | b | 0.5 | 22.4 | 0.0 | 15.9ᶜ |
| Singapore | 100.0 | 1.6 | 3.7 | 1.0 | 60.1 | 27.0 | 3.1 | 3.5 |
| Taiwan | 100.0 | 5.3 | 15.1 | b | 5.0 | 38.5 | b | 11.1 |
| Thailand | 100.0 | 7.9 | 21.0 | 9.6 | b | 21.4 | 0.0 | 39.7ª |

*Source*: Appendix Table 4.A9.

*Notes*:
a Including industrial machinery + equipment.
b Data in the cell have been suppressed to avoid disclosure of data of individual companies.
c Including primary and fabricated metals.

production, such as Jamaica and Malaysia, did not have particularly large affiliate chemicals production.

Large percentage shares in an industry may involve production that is, in total, very small and of little importance to a host country, while smaller shares of larger totals may have larger impacts. The various countries are compared with respect to affiliate production shares of GDP in broad manufacturing industry groups in Table 4.11.

*Table 4.11* Manufacturing gross product of US MOFAs, by industry, as percentage of host country manufacturing GDP, 1994

| Country /Region | Total Mfg | Food and kindred products | Chemicals and allied products | Primary and fabricated metals | Industrial machinery and equipt | Electronic and other electrical equipt | Transport equipt | Other Mfg |
|---|---|---|---|---|---|---|---|---|
| *World* | 0.81 | 0.10 | 0.16 | 0.03 | 0.11 | 0.08 | 0.14 | 0.19 |
| *Developed countries* | 0.82 | 0.10 | 0.16 | 0.03 | 0.12 | 0.07 | 0.15 | 0.20 |
| *Developing countries* | 0.92 | 0.13 | 0.19 | 0.03 | 0.10 | 0.14 | 0.15 | 0.19 |
| Latin America and Caribbean | 1.65 | 0.28 | 0.38 | 0.07 | 0.07 | 0.11 | 0.36 | 0.38 |
| Argentina | 0.82 | 0.35 | 0.23 | 0.02 | 0.02 | 0.02 | 0.01 | 0.17 |
| Brazil | 2.10 | 0.21 | 0.44 | 0.09 | 0.13 | 0.06 | 0.62 | 0.56 |
| Chile | 1.26 | 0.15 | 0.27 | 0.41 | 0.00 | 0.02 | 0.16 | 0.25 |
| Colombia | 1.49 | b | 0.53 | 0.05 | 0.00 | b | b | b |
| Ecuador | 0.48 | 0.22 | 0.14 | 0.02 | 0.00 | b | b | 0.08 |
| Peru | 0.21 | 0.05 | 0.09 | 0.00 | 0.00 | 0.01 | 0.00 | 0.06 |
| Venezuela | 1.71 | 0.37 | 0.37 | 0.07 | 0.03 | b | b | 0.51 |
| Costa Rica | 2.26 | 0.70 | 0.49 | 0.17 | 0.00 | b | 0.00 | b |
| Guatemala | 0.74 | 0.21 | 0.16 | 0.05 | 0.00 | 0.00 | 0.00 | 0.32 |
| Honduras | 4.62 | b | 0.09 | 0.06 | 0.00 | 0.00 | 0.00 | b |
| Mexico | 2.16 | 0.38 | 0.51 | 0.05 | 0.09 | 0.35 | 0.45 | 0.33 |
| Panama | 1.02 | 0.19 | 0.29 | 0.06 | 0.00 | 0.00 | 0.00 | 0.50 |
| Jamaica | 2.25 | 0.00 | 1.09 | 0.00 | 0.00 | 0.00 | 0.00 | 1.16 |
| Africa | 0.25 | 0.07 | 0.09 | 0.02 | 0.01 | 0.00 | 0.00 | 0.05 |
| South Africa | 0.44 | 0.07 | 0.20 | 0.03 | 0.03 | 0.00 | 0.02 | 0.09 |
| Other | 0.17 | 0.07 | 0.04 | 0.02 | 0.00 | 0.00 | 0.00 | 0.03 |
| Middle East | 0.19 | 0.02 | 0.01 | 0.00 | 0.00 | 0.13 | 0.00 | 0.02 |
| Israel | 0.84 | b | 0.04 | 0.00 | 0.00 | 0.60 | 0.00 | b |
| Other | 0.02 | b | 0.01 | 0.00 | 0.00 | 0.00 | 0.00 | b |
| Asia and Pacific | 0.51 | 0.03 | 0.07 | 0.01 | 0.13 | 0.17 | 0.02 | 0.07 |
| China | 0.10 | 0.01 | 0.02 | 0.01 | 0.02 | 0.03 | 0.00 | 0.01 |
| Hong Kong | 1.00 | 0.01 | 0.07 | 0.06 | 0.09 | 0.44 | 0.02 | 0.31 |
| India | 0.07 | b | 0.03 | b | 0.02 | −0.92 | 0.00 | 0.03 |
| Indonesia | 0.22 | 0.01 | 0.15 | 0.00 | b | 0.01 | 0.01 | 0.05[a] |
| Korea, Republic of | 0.20 | 0.02 | 0.03 | 0.00 | 0.01 | 0.06 | 0.01 | 0.06 |
| Malaysia | 2.31 | 0.03 | 0.11 | b | 0.31 | 1.47 | 0.00 | 0.39[c] |
| Philippines | 1.52 | 0.38 | 0.55 | b | 0.01 | 0.34 | 0.00 | 0.24[c] |
| Singapore | 5.41 | 0.08 | 0.20 | 0.06 | 3.25 | 1.46 | 0.17 | 0.19 |
| Taiwan | 0.65 | 0.03 | 0.10 | b | 0.03 | 0.25 | b | 0.07 |
| Thailand | 0.55 | 0.04 | 0.11 | 0.05 | b | 0.12 | 0.00 | 0.22[a] |

*Sources*: Appendix Tables 4.A1 and 4.A9.

*Notes*:
a Including industrial machinery and equipment.
b Data in the cell have been suppressed to avoid disclosure of data of individual companies.
c Including primary and fabricated metals.

Some of the same geographical patterns as in the industry distribution emerge. US affiliate manufacturing production is a much more important part of GDP in Latin America than in Asia except in the two machinery industries. All the cases in which US MOFA production in a machinery group was as much as 0.4 per cent of GDP were in Asia and Israel, with only Mexico, among the Latin American countries, close. Electrical machinery was the industry with the larger US affiliate shares. US MOFA food and chemicals production was of importance mainly in Latin America, but the Philippines followed something like the same pattern. Affiliate metals output was as high as 0.4 per cent only in Chile, presumably associated with the large copper mining operations.

The contrast between Latin America and Asia stands out even more clearly in Table 4.12, which compares US affiliate production with each country's manufacturing production, rather than with total output. In Latin America and in the Philippines, the high shares of US affiliate production in aggregate manufacturing output, over 2 per cent, for example, are all in foods and chemicals and, in Mexico and Brazil, in transport equipment. In Asia and in Israel, the high ratios, some as much as 4 per cent of manufacturing production, are in the two machinery industries.

We can perform a more formal analysis of what attracts US affiliate production of various types by regressing these production and production share measures on a few host country characteristics. Aggregate production in each of the industries is explained almost completely by host country GDP. Size of market is of overwhelming importance in every industry. In the two machinery groups, however, but in no other industries, per capita income is also positively related to the amount of US MOFA production. If a dummy variable is added for Asian countries to test whether the per capita income might be a disguised version of some other advantages of these countries, it is insignificant in the equations for the machinery industries; per capita income, or the skills and education it represents, remains a strong positive influence in attracting these industries. However, the negative influence of per capita income on MOFA production in chemicals disappears, replaced by a negative coefficient on the Asia variable, and the coefficients for that same dummy variable are negative and significant or close to it in the equations for foods, metals, and other manufacturing. It is not clear what these negative dummy variable coefficients represent, possibly lack of natural resources or distance from major consuming markets, more important for low value commodities than for high value electronics.

If we relate industry shares of MOFA production to country GDP alone, the only significant equation is for transport equipment. The higher the GDP, the larger the share of transport equipment in US MOFA production. In no other industry is the size of the market a determinant of the industry's share of production. Per capita income does better in explaining production shares; it is positively related to the production shares of the two machinery industries and negatively related to the production share of chemicals. If the two variables are used together, the degree of explanation does not improve. Country size is never significant, although it comes closest in the case of transport equipment, again suggesting that economies of scale are probably most important in that industry. High per capita income is positively related to the share of both machinery industries, negatively and significantly related

*Table 4.12* Manufacturing gross product of US MOFAs as percentage of host country manufacturing GDP, 1994 (per cent)

| Country/Region | Total Mfg | Food and kindred products | Chemicals and allied products | Primary and fabricated metals | Industrial machinery and equipt | Electronic and other electrical equipt | Transport equipt | Other Mfg |
|---|---|---|---|---|---|---|---|---|
| *Developing countries* | 2.88 | 0.40 | 0.59 | 0.11 | 0.30 | 0.44 | 0.46 | 0.58 |
| *Latin America and Caribbean* | 6.74 | 1.16 | 1.55 | 0.27 | 0.28 | 0.45 | 1.49 | 1.55 |
| Argentina | 3.75 | 1.61 | 1.05 | 0.11 | 0.08 | 0.08 | 0.05 | 0.77 |
| Brazil | 9.47 | 0.94 | 1.98 | 0.39 | 0.58 | 0.27 | 2.79 | 2.53 |
| Chile | 5.42 | 0.64 | 1.18 | 1.75 | 0.00 | 0.07 | 0.68 | 1.10 |
| Colombia | 4.86 | b | 1.71 | 0.15 | −0.01 | b | b | b |
| Ecuador | 2.22 | 1.03 | 0.67 | 0.11 | 0.00 | b | 0.00 | 0.39 |
| Peru | 1.03 | 0.24 | 0.46 | 0.02 | 0.00 | 0.03 | 0.00 | 0.28 |
| Venezuela | 11.08 | 2.37 | 2.37 | 0.45 | 0.21 | b | b | 3.34 |
| Costa Rica | 11.74 | 3.64 | 2.57 | 0.88 | 0.00 | b | 0.00 | b |
| Guatemala | 7.05 | 2.00 | 1.56 | 0.45 | 0.00 | 0.00 | 0.00 | 3.04 |
| Honduras | 30.29 | b | 0.62 | 0.41 | 0.00 | 0.00 | 0.00 | b |
| Mexico | 10.93 | 1.94 | 2.59 | 0.24 | 0.44 | 1.76 | 2.29 | 1.66 |
| Panama | 11.04 | 2.05 | 3.16 | 0.63 | 0.00 | 0.00 | 0.00 | 5.36 |
| Jamaica | 13.32 | 0.00 | 6.45 | 0.00 | 0.00 | 0.00 | 0.00 | 6.87 |
| Trinidad and Tobago | 0.00 | 0.00 | 0.00 | 0.00 | 0.00 | 0.00 | 0.00 | 0.00 |
| *Africa* | 2.35 | 0.67 | 0.81 | 0.23 | 0.09 | 0.03 | 0.05 | 0.48 |
| South Africa | 2.12 | 0.34 | 0.97 | 0.13 | 0.14 | 0.02 | 0.08 | 0.45 |
| Other | 2.66 | 1.10 | 0.61 | 0.37 | 0.02 | 0.04 | 0.00 | 0.51 |
| *Middle East* | 2.14 | 0.27 | 0.14 | 0.02 | 0.02 | 1.44 | 0.01 | 0.25 |
| Israel | 6.37 | b | 0.28 | 0.03 | 0.01 | 4.56 | 0.00 | b |
| Other | 0.22 | b | 0.07 | 0.02 | 0.02 | 0.02 | 0.00 | b |
| *Asia and Pacific* | 1.26 | 0.07 | 0.18 | 0.03 | 0.33 | 0.43 | 0.06 | 0.18 |
| China | 0.25 | 0.01 | 0.06 | 0.02 | 0.04 | 0.09 | 0.00 | 0.03 |
| Hong Kong | 9.50 | 0.14 | 0.63 | 0.55 | 0.90 | 4.22 | 0.15 | 2.92 |
| India | 0.46 | b | 0.18 | b | 0.09 | 0.00 | 0.00 | 0.17 |
| Indonesia | 0.93 | 0.04 | 0.61 | 0.02 | b | 0.05 | 0.02 | 0.20c |
| Korea, Republic of | 0.70 | 0.08 | 0.12 | 0.01 | 0.05 | 0.21 | 0.02 | 0.22 |
| Malaysia | 10.78 | 0.13 | 0.53 | b | 1.45 | 6.86 | 0.00 | 1.81c |
| Philippines | 6.50 | 1.61 | 2.37 | b | 0.03 | 1.46 | 0.00 | 1.03c |
| Singapore | 20.06 | 0.31 | 0.74 | 0.20 | 12.06 | 5.42 | 0.62 | 0.70 |
| Taiwan | 2.14 | 0.11 | 0.32 | 0.05 | 0.11 | 0.82 | 0.49 | 0.24 |
| Thailand | 1.91 | 0.15 | 0.40 | 0.18 | b | 0.41 | 0.00 | 0.76c |

*Sources*: Appendix Tables 4.A3 and 4.A9.

*Notes*:

a Including industrial machinery and equipment.

b Data in the cell have been suppressed to avoid disclosure of data of individual companies.

c Including Primary and fabricated metals.

to the share of chemicals, and negatively, but not significantly related to the shares of foods and 'other manufacturing', suggesting that host country skill or education levels are most important for attracting investment in the machinery industries, and wage levels important in chemicals and possibly the other two. Adding an Asia dummy variable to these share equations improves most, though not all, of them. The dummy variable is positive for both machinery groups and significant for industrial machinery without eliminating the positive and significant coefficients for per capita income.

The shares of US MOFA production in total GDP are not explained at all by market size and only those for the two machinery industries are explained significantly by per capita income. The higher the per capita income, the larger the US MOFA production in these two industries, relative to GDP, either because the high skill or education levels attract this production or because US MOFA production in these industries raises per capita income. In this case, adding an Asia dummy does not affect the relationships with per capita income and the dummy variable is never significant.

What stands out in these equations is the strong association between per capita GDP and US MOFA production in the two machinery industries, either because the factors that produce high incomes make a country a good location for machinery production or because such production tends to raise incomes. That distinction is explored in the next section. Another point that stands out is the lack of any evidence that low per capita income attracts affiliate production in any industry. The apparent relationship for the chemicals industry seems to reflect instead the disadvantage of Asia as a production location for this industry.

## Impacts of FDI on developing host countries

Much of the early theoretical literature on the effects of inward direct investment on host countries produced rather pessimistic conclusions (e.g. MacDougall, 1960). Direct investment was thought of mainly as a flow of capital, possibly replacing local capital or possibly representing marginal additions to the host country's capital stock, followed by the necessity of financing dividends and interest, and possibly repatriation of capital. Later, a more favourable appraisal resulted from the realization that direct investment represented something very different from a simple flow of capital and could take place even without any capital flow, although that was not the typical case. The distinctive feature of direct, as opposed to portfolio, investment is that direct investment involves the exploitation of the investor's firm-specific assets, such as superior technology. The investor brings to the host country not only capital but, more crucially, superior technology in manufacturing marketing or other aspects of the enterprise that could not or would not be transferred without production in the host country.

### Impacts on trade

One of the places where an impact is most likely to be observed is in the amount and composition of a host country's trade. In the mid-1960s, US manufacturing affiliates in developing countries existed almost entirely to serve their host country markets. Many of them operated in Latin America behind various types of protection and were too small or too inefficient to compete in export markets. While US affiliates in developed countries exported a fifth of their output in 1966, those in developing countries exported only 8 per cent. Since that time these developing-country affiliates have been transformed. While the export propensities of affiliates in developed countries have grown, those of developing-country affiliates have increased much faster, so that since 1989 the export propensities of the two groups have been very similar. US affiliates in both sets of countries export well over a third of the value of their output.

There have been, for as long as we have records, large regional differences in the market orientation of affiliates. Those in Asia were, even in 1966, export-oriented to a much greater degree than those in Latin America, partly because some of the Asian host countries were small and probably also because they did not have such inward-looking economic policies. Affiliates in the NIEs were the extreme cases, producing almost entirely for export in 1977, the first year for which we can observe them separately. Over the next decade and a half, while US manufacturing affiliates in the developed countries, in Latin America and in Asia outside the NIEs were shifting toward more exporting, those in the NIEs were shifting towards own domestic markets, although they were still exporting two thirds of their output in 1993.

Despite their increasing export orientation, US affiliates have never accounted for the majority of host country exports. For all host countries combined, the US (majority-owned) affiliates accounted for 6.5 per cent of exports in 1966 and 8.5 per cent in 1993. The share increased in every group of countries, suggesting that the MNEs were leading their host countries towards greater participation in exporting to world markets. The advantages that MNEs bring to their host countries are their access to their home markets and also their sunk costs in market research, sales and service facilities, advertising, and other means of selling in diverse markets. They can also bring diversification in markets to countries in which local firms, sometimes under the umbrella of protection, have been focused mainly on their own countries' internal markets.

The anatomy of the role of US affiliates can be seen more clearly in the events that followed the Latin American debt crisis in 1982. US affiliates in heavily indebted countries such as Brazil, Chile and Mexico switched the geographical composition of their sales from host country markets to export markets much more quickly and more completely than other firms, presumably because they already had links to other markets. At first, much of the switch involved virtually cutting off sales in the host countries rather than increasing their export sales, a measure that may not have been available to local firms, more dependent on their local customers. Furthermore, the US affiliates may have been reluctant to sell locally for depreciating host country currencies, while local firms were less affected by currency depreciation because both their suppliers and their owners were local to a greater degree. After four or five years, the US affiliates were increasing both their local and their export sales. Their export propensities were not only higher than those of local firms but had risen more after the debt crisis. Thus, the flexibility of the US affiliates helped to bring about the market switching for the host countries that was an important part of the recovery from the debt crisis (Blomström and Lipsey, 1993).

Another piece of evidence for the role of US MNEs in enhancing the ability of developing countries to break into export markets is provided by the experience of Asian affiliates of US firms. An examination of the determinants of export market penetration by these affiliates found that, given the size of a market, as measured by population or real GDP, and given the market's total imports of the products of the affiliate's industry, the larger the export of an Asian affiliate's parent to its affiliates in that market, the greater the exports by the Asian affiliate of that parent to that same market. In other words, an affiliate's ability to export to a market was enhanced by strong connections between its parent and the parent's affiliates in that market. Parent exports to a market conducted at arm's length, rather than through affiliates, tended to compete with, rather than promote, exports by these Asian affiliates, although the effect was much smaller than the promotion effect of exports to affiliates. The effect of MNE connections was particularly strong in Asian affiliate exports of electronic components and accessories, as might be expected given the ability of MNEs to break up the production of these into segments suitable for different locations (Lipsey, 1995).

An obvious question is why this geographical allocation of production could not take place through trade without the involvement of MNEs. Of course, most of the international allocation of production does take place through trade. But there are aspects of production that must take place under a single management, especially if production involves inter-related segments with distinctive characteristics that must match, or if production involves proprietary information that must be shared among the segments, or if it involves the use of knowledge gained from the firm's R&D. In the absence of the possibility of internationalized production, all these segments must take place within one country, even if some would be much more economically performed elsewhere. A high-tech firm in a labour-scarce economy would have to perform the labour-intensive portions of its production at home even though labour is expensive there. A firm from a country with few well-trained scientists would have to perform its R&D at home even though it might be cheaper and more efficient in a more educated country. What internationalization does is to extend the boundaries of the operation of comparative advantage so that these activities, otherwise confined to a single country, usually a firm's home country, can be located on the basis of geographical comparative advantage. A semiconductor company can produce wafers at home and assemble the semiconductors where labour is cheap. A producing company in a developing country can place its research and development operations in a country with research universities and laboratories. For developing countries and developed countries it adds to the 'suitable' activities in each and reduces the 'unsuitable' activities.

A source of influence on host country trade that is rarely discussed is R&D conducted by MNEs within developing countries. The reason for that neglect is that these MNE affiliates are far less involved in R&D than their parents in the United States or their siblings in developed countries. For example, in 1982, developing country manufacturing affiliates were about a third as R&D-intensive as the corresponding affiliates in developed countries and the latter were less than half as R&D-intensive as parents. It is therefore not surprising that, across industries, the share of US affiliates in host country exports was more strongly correlated with parent R&D intensity than with that of affiliates. Despite the low level of affiliate R&D activity, compared with that of parents, US affiliates account for significant shares of total R&D in some of the developing countries where they operate. Therefore, it is not inconceivable that they would have an impact on the composition of host country exports. If we divide developing country exports into high, medium, and low-tech products, we find that the higher the R&D intensity of US affiliates in a country, the higher the share of high-tech products in the country's exports, even after we take account of country characteristics that could explain export composition, such as per capita real income and the proportion of professional and technical workers in the labour force (Lipsey, Blomström and Kravis, 1990). Another indication of the technology element in US direct investment in developing countries is given by the industrial composition of affiliate exports. As compared with developed countries, developing countries have comparative advantages in food industries and in 'other manufacturing', which includes textiles, apparel, and leather goods industries as well as toys. The distribution of exports from Japanese affiliates in developing countries is more like that of the developing countries as a group than that of Japan. However, the exports of US affiliates are more like those of the United States, with 70 per cent in machinery and transport equipment, the manufacturing industry groups in which R&D and technological advances are concentrated (Lipsey, Blomström and Kravis, 1990). Thus, the chance of technological progress from the US-owned production may be greater than from the Japanese-owned production in developing host countries.

### Impacts on productivity, technology, and growth rates

A different channel of MNE impacts on host countries in what are called 'spillovers' from foreign-owned operations to the domestic economy of the host country. That term is usually construed to refer to effects on locally owned firms that continue in operation. It omits the effects on the local economy that take place entirely within the foreign-owned operations, leaving domestically owned operations unaffected, and it omits the effects on the productivity of the host economy that come from driving out inefficient local firms. Often the available data cannot make these distinctions, but if one is primarily concerned with the growth of output and productivity in the geographically defined local economy, these distinctions may not be essential.

One study of Mexico found that, across industries, the higher the degree of foreign ownership, the faster the growth in labour productivity and the faster the rate at which productivity in Mexican-owned firms caught up to or gained on the productivity of the MNE's operations in Mexico and to the productivity of the corresponding industries in the United States (Blomström and Wolff, 1994). A less favourable view of the effect of foreign operations on locally owned firms in several developing countries comes out of a series of studies based on establishment microdata. They all show that foreign firms have higher levels of labour productivity and of total factor productivity than locally owned firms but that the presence of the foreign firms did not tend to raise the productivity of local firms in the same industries and, in some cases, seemed to reduce it. That could be the case, for example, if the foreign firms hired away the best workers from their locally owned rivals. Even those cases are consistent with host country gains in productivity from the entry of foreign firms, and presumably gains to consumers, but perhaps some of these come at the expense of local entrepreneurs (Aitken and Harrison, 1993; Aitken, Harrison, and Lipsey, 1996; Haddad and Harrison, 1993; Harrison, 1992; Harrison, 1996; and Luttmer and Oks, 1993). All of these studies examine effects of foreign entry on wages as well as on productivity, and pretty unanimously conclude that foreign firms pay higher wages than local firms and tend to increase the overall level of wages where they operate, but that the overall impact comes solely from the foreign establishments, not from any spillover of their wage levels to those paid by local firms. In fact, in some cases, the higher wages in foreign plants are partly offset by lower wages in locally owned plants. The relationship between foreign and local plants contrasts with that in the United States, where a higher degree of foreign ownership in an industry in a state is associated with higher wages and productivity not only overall but in US-owned plants as well. That spillover of wage levels and productivity to locally owned plants is not observed in most developing countries, perhaps because labour markets and product markets are less competitive there.

One technological aspect of direct investment that is often overlooked is the resulting improvement in the performance of service industries, particularly business services, which are inputs to manufacturing and other goods industries. While few services are R&D-intensive, many are human capital-intensive, and it is the technology embodied in human capital that is the basis for FDI in most service industries. While US affiliates in goods industries pay much lower wages, on average, than their parents, presumably because skill levels are lower as well as because labour of any skill is cheaper, affiliates in service industries pay more than manufacturing affiliates in the same countries and at a level much closer to that in the United States. In some industries, US affiliates in developing countries pay higher wages than those in developed countries and even higher wages than in their parent companies, a strong hint that they have brought high-skill operations to developing countries. These technologically advanced services are, in turn, extremely important to the competitiveness and export success of the country's goods industries (UNCTAD, 1994a).

There have been a number of macro-economic studies pointing to the benefits of inward direct investment. The ratio of inward FDI to total output was found to have been a positive influence on long-term growth in Blomström, Lipsey and Zejan (1994) and over shorter periods as well in Blomström, Lipsey and Zejan (1996). And Romer (1993) has placed considerable weight on the role of direct investment as a vehicle for reducing the 'idea gap' between developed and developing countries.

> successful development... requires a mechanism for ensuring adequate flows of the large quantity of disembodied ideas that are used in production. The government of a poor country can therefore help its residents by creating an economic environment that offers an adequate reward to multinational corporations when they bring ideas from the rest of the world and put them to use with domestic resources.
>
> (Romer, 1993: 548)

An association between high host country per capita income and US firms' production of machinery was noted above and ascribed to the effect of local skills and capital in attracting such production. An alternative interpretation would be that the foreign investment produced the high per capita incomes. A test of that relationship by relating growth in per capita income from 1977 to 1994 to the 1977 shares of affiliate production in total and manufacturing output produced no significant equations and the coefficients on foreign production shares were negative, although not significant.

## Summary

Despite the high volume of academic and popular literature on globalization or internationalization, there has been no recent trend toward internationalization of US output, in the sense of an increase in production owned by US firms outside the US relative to production in the US. If anything, the trend has been down since the mid-1970s. Manufacturing has been much more internationalized in this sense than US production in general and the petroleum industry even more so, with US-owned production outside the US as large as production in the US. Outside those two sectors, US production internationalization has been negligible.

117

Developing countries' share of internationalized US production fell by half from 1977 to 1991 and then recovered somewhat, but remains far below earlier levels, reflecting the shift in US FDI from resource-oriented to market-oriented investment. The decline in the developing countries' share of petroleum industry investment was even sharper but in manufacturing there was a small shift towards the developing countries. However, internationalized manufacturing production by US MNEs remains concentrated in developed countries and will almost certainly remain so. Thus, neither any trend toward internationalization nor any prospective shift towards developing countries as a group by US firms are likely to be major sources of developing country economic growth.

American firms' total internationalized output and output in manufacturing has declined relative to total production and manufacturing output in developing countries as a whole and in most individual countries except in the Asia-Pacific region. There it has increased in importance in the most successful and most open of the countries. The exceptions to the general decline in US firms' shares are almost all countries with relatively open economies or economies that have reformed their economic regimes to encourage foreign investment, including, Latin America, Chile and Mexico. Thus there are indications that the role of internationalized production is to a large extent under the control of host country governments and their economic policies.

One way in which internationalized production affects host countries is by the connections the foreign firms form with local firms. The less self-sufficient the foreign operations, the greater their linkages to local suppliers, unless they depend on unrelated suppliers in other countries. In most regions, the share of manufacturing sales produced within the host countries has been fairly stable, but it has fallen sharply in the Asia-Pacific region and risen a little in Latin America. The share of sales produced within host countries is much higher in Latin America than in Asia, possibly reflecting the protectionist policies of the former region.

Manufacturing affiliates in Latin America have become more dependent on parents over time, while those in Asia have become less dependent. Another source of supply outside local firms is related affiliates in the host country and elsewhere, leaving for local suppliers, and unrelated foreign suppliers, purchases equal to about 40 per cent of sales, a figure that has not changed substantially. Thus there is little indication of any strong trend by US affiliates towards increasing self-sufficiency or towards increasing integration into local host country economies.

Within manufacturing, different developed countries attract different types of industries. Electrical machinery and 'other manufacturing' were much more important in US affiliate production in Asia, while foods, chemicals and transport equipment were more important in Latin America. The level of affiliate production in a country is explained to a large extent by the size of the economy but both industrial and electrical machinery production are drawn to a host country by relatively high per capita income, presumably reflecting host country physical and human capital stocks. Asian locations seem to be relatively unfavourable for chemicals production and large host country market size is a favourable factor mainly for production of transport equipment. Low per capita income, presumably associated with low wage levels, does not seem to attract affiliate production in any of these broad industry groups.

The impacts of US FDI can be grouped in three categories. US affiliates have been important promoters of host country exports, especially in countries with export-oriented and relatively open economic policies. The affiliates' international market connections make them more adept than local firms at switching between host country local and foreign markets as market conditions dictate. These international market connections also enable host country affiliates to export more than other firms to markets where their parent firms have major market presence. R&D within the host country foreign-owned affiliates encourages host country export of high-tech products beyond the level expected from other host country characteristics, such as their levels of per capita income and skilled labour force.

Internationalized manufacturing production appears to raise wages and productivity in host country industries. Whether the effects extend beyond the foreign affiliates seems to vary among countries, possibly because labour market and other economic policies differ greatly. Presumably part of the effect is to eliminate inefficient locally-owned operations by competition, but there is also some evidence of effects through improvements in remaining local enterprises.

Several time series studies have found a positive impact of inward FDI on host country rates of economic growth, taking account of various other determinants of growth rates, and such an effect is sometimes included in models of economic growth. These studies have difficult problems disentangling effects of investment on growth from effects of growth on investment. A crude test for US-owned production in the countries included here between 1977 ratios of affiliate production to total output and growth in per capita income from 1977 to 1994 did not give any support for a causal connection.

# APPENDIX TABLE 4.A1

## Nominal GDP of developing countries by region and country, 1977–94 (millions of US$)

| Country/Region | 1977 | 1982 | 1989 | 1991 | 1992 | 1993 | 1994 |
|---|---|---|---|---|---|---|---|
| | | | | Year | | | |
| Developing countries total | 1,466,803 | 2,443,791 | 3,155,445 | 3,683,984 | 4,005,304 | 4,277,134 | 4,802,117 |
| Latin America and Caribbean | 476,608 | 817,108 | 993,085 | 1,185,943 | 1,305,548 | 1,473,914 | 1,675,261 |
| Argentina | 56,781 | 84,307 | 76,637 | 189,720 | 228,779 | 257,841 | 281,922 |
| Brazil | 176,084 | 281,544 | 448,764 | 405,770 | 409,167 | 507,353 | 639,562 |
| Chile | 13,366 | 24,340 | 28,182 | 34,396 | 42,748 | 45,639 | 52,181 |
| Colombia | 19,471 | 38,967 | 39,538 | 41,700 | 48,699 | 54,076 | 64,371 |
| Ecuador | 6,655 | 12,447 | 9,823 | 11,752 | 12,656 | 14,304 | 16,557 |
| Peru | 14,231 | 24,817 | 26,168 | 29,310 | 30,193 | 41,106 | 50,385 |
| Venezuela | 43,756 | 79,256 | 43,551 | 53,441 | 60,423 | 59,995 | 58,882 |
| Costa Rica | 3,072 | 2,607 | 5,226 | 5,637 | 6,722 | 7,556 | 8,284 |
| Guatemala | 5,481 | 8,717 | 8,411 | 9,406 | 10,441 | 11,309 | 12,855 |
| Honduras | 1,670 | 2,904 | 3,479 | 2,999 | 3,328 | 3,357 | 3,161 |
| Mexico | 87,343 | 173,789 | 206,225 | 290,535 | 334,356 | 365,857 | 375,465 |
| Panama | 2,077 | 4,325 | 4,639 | 5,496 | 6,015 | 6,565 | 6,852 |
| Jamaica | 3,250 | 2,948 | 4,063 | 3,706 | 3,355 | 4,192 | 4,231 |
| Trinidad and Tobago | 3,139 | 8,140 | 4,323 | 5,308 | 5,440 | 4,669 | 4,795 |
| Africa | 224,086 | 367,866 | 384,421 | 412,936 | 425,345 | 420,122 | 422,254 |
| South Africa | 38,252 | 74,166 | 91,753 | 112,302 | 119,249 | 117,228 | 124,726 |
| Other | 185,834 | 293,700 | 292,668 | 300,634 | 306,096 | 302,894 | 297,528 |
| Middle East | 203,603 | 354,868 | 313,785 | 360,968 | 362,216 | 362,998 | 378,835 |
| Israel | 15,098 | 26,222 | 46,827 | 63,032 | 69,763 | 69,739 | 78,401 |
| Other | 188,505 | 328,646 | 266,958 | 297,936 | 292,453 | 293,259 | 300,434 |
| Asia and Pacific | 562,506 | 903,949 | 1,464,154 | 1,724,137 | 1,912,195 | 2,020,101 | 2,325,767 |
| China | 169,428 | 197,786 | 323,838 | 377,439 | 418,462 | 425,542 | 508,180 |
| Hong Kong | 15,599 | 31,712 | 67,162 | 86,024 | 100,681 | 116,034 | 131,881 |
| India | 112,192 | 185,019 | 274,152 | 251,258 | 242,355 | 250,721 | 291,054 |
| Indonesia | 48,396 | 94,715 | 94,449 | 116,622 | 128,025 | 158,009 | 174,638 |
| Korea, Republic of | 37,064 | 75,620 | 222,152 | 294,175 | 307,938 | 330,831 | 376,505 |
| Malaysia | 13,724 | 26,796 | 37,872 | 47,111 | 58,014 | 63,338 | 70,634 |
| Philippines | 19,649 | 37,140 | 42,575 | 45,417 | 52,977 | 54,389 | 63,876 |
| Singapore | 6,575 | 15,266 | 29,146 | 42,277 | 49,501 | 57,153 | 68,949 |
| Taiwan | 21,816 | 47,606 | 149,158 | 179,779 | 212,150 | 222,385 | 240,744 |
| Thailand | 19,779 | 36,590 | 72,251 | 98,742 | 111,546 | 125,222 | 143,209 |

Sources: World Bank (1995); China, Republic of (1995), exchange rates from IMF (1995). The 1994 totals for the world and for developed countries were 25,299,043 and 19,907,748.

# APPENDIX TABLE 4.A2

## Nominal gross product of US non-bank MOFAs[a] in developing countries by region and country, 1977–95 (millions of US$)

| Country/Region | Year | | | | | | | |
|---|---|---|---|---|---|---|---|---|
| | 1977 | 1982 | 1989 | 1991 | 1992 | 1993 | 1994 | 1995 |
| Developing countries total | 52,339 | 56,940 | 56,138 | 57,780 | 65,992 | 73,689 | 79,217 | 93,793 |
| **Latin America and Caribbean** | 16,036 | 27,939 | 29,601 | 28,464 | 33,635 | 38,995 | 41,667 | 45,820 |
| Argentina | 1,449 | 2,902 | 1,577 | 3,363 | 3,748 | 4,174 | 4,245 | 4,585 |
| Brazil | 6,485 | 11,199 | 16,618 | 11,514 | 14,457 | 17,661 | 16,826 | 18,587 |
| Chile | 162 | 468 | 681 | 926 | 1,310 | 1,222 | 1,717 | 2,266 |
| Colombia | 532 | 1,361 | 1,150 | 1,278 | 1,545 | 1,626 | 1,830 | 2,134 |
| Ecuador | 307 | 516 | 272 | 327 | 169 | 131 | 220 | 182 |
| Peru | 404 | 1,116 | 397 | 340 | 361 | 460 | 660 | 1,047 |
| Venezuela | 1,370 | 2,394 | 736 | 1,080 | 1,309 | 1,473 | 1,575 | 2,230 |
| Costa Rica | 115 | 163 | 208 | 192 | 413 | 383 | 503 | 552 |
| Guatemala | 156 | 276 | 158 | 238 | 232 | 206 | 232 | 235 |
| Honduras | 142 | 251 | 287 | 276 | 270 | 212 | 286 | 314 |
| Mexico | 2,050 | 3,561 | 4,883 | 7,585 | 8,664 | 8,785 | 9,849 | 7,733 |
| Panama | 289 | 433 | 530 | 561 | 540 | 286 | 351 | 442 |
| Jamaica | 370 | 403 | 455 | 334 | 295 | 262 | 264 | 359 |
| Trinidad and Tobago | – | – | 497 | 642 | 570 | 594 | 330 | 666 |
| **Africa** | 6,703 | 7,725 | 4,598 | 5,322 | 5,439 | 4,985 | 5,411 | 6,641 |
| South Africa | 1,317 | 2,330 | 701 | 752 | 802 | 757 | 908 | 1,109 |
| Other | 5,386 | 5,395 | 3,897 | 4,570 | 4,637 | 5,227 | 4,503 | 5,532 |
| **Middle East** | 22,260 | 8,112 | 4,891 | 2,862 | 3,657 | 3,078 | 3,071 | 3,839 |
| Israel | 225 | 280 | 359 | 632 | 705 | 775 | 934 | 1.071 |
| Other | 22,035 | 7,832 | 4,532 | 2,230 | 2,952 | 2,303 | 2,137 | 2,768 |
| **Asia and Pacific** | 7,340 | 13,164 | 17,048 | 21,132 | 23,261 | 25,632 | 29,068 | 37,493 |
| China | 2 | 7 | 8 | 211 | 359 | 333 | 678 | 1,076 |
| Hong Kong | 542 | 959 | 2,926 | 3,192 | 3,485 | 4,205 | 4,900 | 6,700 |
| India | 210 | 229 | 157 | 123 | 121 | 176 | 232 | 459 |
| Indonesia | 4,661 | 6,317 | 3,999 | 5,031 | 5,100 | 4,874 | 4,649 | 5,850 |
| Korea, Republic of | 79 | 219 | 726 | 1,031 | 1,079 | 1,122 | 1,452 | 2,002 |
| Malaysia | 333 | 1,691 | 1,749 | 2,016 | 3,117 | 3,127 | 3,579 | 4,283 |
| Philippines | 549 | 1,074 | 1,006 | 1,189 | 1,413 | 1,674 | 1,803 | 2,046 |
| Singapore | 400 | 1,109 | 2,353 | 3,333 | 3,298 | 4,761 | 5,750 | 7,675 |
| Taiwan | 260 | 616 | 1,938 | 2,395 | 2,678 | 2,462 | 2,810 | 3,347 |
| Thailand | 254 | 657 | 1,815 | 2,203 | 2,130 | 2,405 | 2,644 | 3,414 |

*Sources*: Mataloni and Goldberg (1994), Mataloni (1995), (1997) and Mataloni and Fahim-Nader (1996).
*Note*:
a Non-bank majority-owned foreign affiliates of non-bank parents.

# APPENDIX TABLE 4.A3

## Manufacturing nominal GDP in developing countries by region and country, 1977–94 (millions of US$)

| Country/Region | Year | | | | | | |
|---|---|---|---|---|---|---|---|
| | 1977 | 1982 | 1989 | 1991 | 1992 | 1993 | 1994 |
| Developing countries total[a] | 270,910 | 448,146 | 907,836 | – | – | – | 1,435,357[e] |
| Latin America and Caribbean[a] | 114,253 | 180,710 | 242,796 | – | – | – | 409,579[e] |
| Argentina | 20,779 | 26,439 | 23,719 | 46,266 | 50,009 | – | 61,626[e] |
| Brazil | 45,530 | 87,481 | 120,845 | 90,062 | – | – | 141,953[e] |
| Chile[a] | 2,906 | 4,586 | 6,547 | – | – | – | 12,122[e] |
| Colombia | 4,642 | 8,269 | 8,259 | 8,393 | 9,618 | – | 19,750[e] |
| Ecuador | 1,197 | 2,212 | 2,071 | 2,441 | 2,790 | 3,110 | 3,600[e] |
| Peru (b) | 2,966 | – | – | – | – | 8,488 | 10,228 |
| Venezuela | 6,683 | 12,609 | 7,033 | 8,232 | 8,838 | 8,669 | 9,068 |
| Costa Rica | 583 | 530 | 1,064 | 1,123 | 1,380 | 1,488 | 1,593 |
| Guatemala[c] | 664 | 1,005 | 871 | 986 | – | – | 1,348[e] |
| Honduras | 222 | 384 | 468 | 435 | 510 | 512 | 482[e] |
| Mexico | 19,605 | 36,044 | 50,412 | 63,784 | 67,221 | 72,225 | 74,122[e] |
| Panama | 234 | 394 | 354 | 452 | 502 | 535 | 634 |
| Jamaica | 599 | 530 | 797 | 697 | 620 | 706 | 713[e] |
| Trinidad and Tobago | 428 | 562 | 414 | 485 | 496 | 425 | 444 |
| Africa[a] | 20,952 | 33,104 | 41,241 | – | – | – | 45,299[e] |
| South Africa | 7,961 | 16,872 | 20,669 | 25,061 | 26,052 | 24,847 | 25,879 |
| Other | 12,991 | 16,232 | 20,572 | – | – | – | 19,420 |
| Middle East[a] | 13,081 | 23,260 | 27,315 | – | – | – | 32,978[e] |
| Israel[c] | 2,984 | 3,870 | 6,152 | – | – | – | 10,300[e] |
| Other | 10,097 | 19,390 | 21,163 | – | – | – | 22,678 |
| Asia and Pacific | 122,625 | 211,073 | 596,486 | – | – | – | 947,501[e] |
| China | 53,584 | 75,802 | 120,352 | 127,984 | 147,302 | 160,645 | 194,048 |
| Hong Kong | 3,494 | 6,272 | 12,330 | 12,511 | 12,888 | 12,190 | 13,855[e] |
| India | 17,162 | 29,158 | 46,256 | 39,446 | 38,382 | 38,981 | 46,787 |
| Indonesia | 5,095 | 11,312 | 17,131 | 24,440 | 27,854 | 35,245 | 41,749 |
| Korea, Republic of | 10,214 | 21,360 | 68,884 | 83,899 | 85,454 | 95,610 | 108,810[e] |
| Malaysia | 2,524 | 4,890 | 8,498 | 11,487 | 12,829 | 13,581 | 15,145[e] |
| Philippines | 4,997 | 9,322 | 10,589 | 11,497 | 12,811 | 12,891 | 14,905 |
| Singapore | 1,632 | 3,810 | 8,558 | 12,373 | 13,034 | 15,234 | 18,583 |
| Taiwan[d] | 7,464 | 16,761 | 51,541 | 59,937 | 67,256 | 67,776 | 73,371[e] |
| Thailand | 3,990 | 7,802 | 19,337 | 28,064 | 31,185 | 35,661 | 40,926 |

*Source*: United Nations, 1993; Republic of China, 1996.

*Notes*:

a  For 1989 extrapolated as the proportion of MFG in GDP from UN (1993).

b  For 1982 and 1989 extrapolated as the proportion of MFG in GDP from UN (1993).

c  For 1977–91 extrapolated as the proportion of MFG in GDP from UN (1993).

d  Republic of China (1996), p. 38.

e  Assuming same ratio of manufacturing gross product to GDP as in latest available year.

# APPENDIX TABLE 4.A4

## Gross product of US manufacturing MOFAs in developing countries by region and country, 1977–95 (millions of US$)

| Country/Region | Year | | | | | | | |
|---|---|---|---|---|---|---|---|---|
| | 1977 | 1982 | 1989 | 1991 | 1992 | 1993 | 1994 | 1995 |
| Developing countries total | 11,933 | 21,862 | 28,903 | 29,544 | 34,964 | 38,097 | 41,333 | 46,953 |
| Latin America and Caribbean | 9,533 | 17,531 | 21,664 | 21,004 | 25,150 | 27,317 | 27,615 | 29,271 |
| Argentina | 945 | 1,859 | 973 | 1,851 | 2,279 | 2,662 | 2,311 | 2,455 |
| Brazil | 5,169 | 9,572 | 14,167 | 9,887 | 12,369 | 13,701 | 13,450 | 15,359 |
| Chile | 62 | 122 | 364 | 325 | 369 | 430 | 657 | 785 |
| Colombia | 320 | 720 | 650 | 641 | 740 | 820 | 959 | 1,101 |
| Ecuador | 35 | 72 | 37 | 33 | 35 | 42 | 80 | 70 |
| Peru | 74 | 108 | 90 | 63 | 71 | 76 | 105 | 152 |
| Venezuela | 745 | 1,401 | 509 | 711 | 952 | 1.083 | 1,005 | 1,361 |
| Costa Rica | 57 | 68 | 99 | 135 | 145 | 154 | 187 | 174 |
| Guatemala | 61 | 74 | 69 | 69 | 79 | 91 | 95 | 96 |
| Honduras | 33 | 82 | 105 | 119 | 131 | 77 | 146 | 141 |
| Mexico | 1,646 | 2,879 | 4,123 | 6,521 | 7,419 | 7,711 | 8,104 | 6,596 |
| Panama | 26 | 45 | 182 | 182 | 201 | 64 | 70 | 93 |
| Jamaica | 18 | 119 | 85 | 103 | 60 | 54 | 95 | 117 |
| Trinidad and Tobago | – | 27 | 16 | 11 | 12 | 11 | −4 | 88 |
| Africa | 802 | 1,345 | 883 | 888 | 972 | 887 | 1,065 | 1,303 |
| South Africa | 546 | 1,011 | 441 | 458 | 492 | 413 | 548 | 723 |
| Other | 256 | 334 | 442 | 430 | 480 | 474 | 517 | 580 |
| Middle East | 103 | 187 | 195 | 384 | 457 | 534 | 706 | 835 |
| Israel | 84 | 124 | 191 | 359 | 431 | 501 | 656 | 737 |
| Other | 19 | 63 | 4 | 25 | 26 | 33 | 50 | 98 |
| Asia and Pacific | 1,495 | 2,799 | 6,161 | 7,268 | 8,385 | 9,359 | 11,947 | 15,544 |
| China | 2 | 4 | 36 | 77 | 122 | 228 | 488 | 957 |
| Hong Kong | 199 | 246 | 751 | 812 | 920 | 926 | 1,316 | 2,147 |
| India | 205 | 209 | 161 | 130 | 122 | 179 | 217 | 351 |
| Indonesia | 106 | 146 | 100 | 114 | 145 | 171 | 390 | 450 |
| Korea, Republic of | 59 | 119 | 463 | 593 | 631 | 671 | 762 | 930 |
| Malaysia | 122 | 373 | 477 | 813 | 1,209 | 1,125 | 1.632 | 1,913 |
| Philippines | 278 | 447 | 625 | 688 | 848 | 926 | 969 | 1,173 |
| Singapore | 210 | 570 | 1,453 | 1,924 | 2,117 | 3,118 | 3,727 | 4,706 |
| Taiwan | 224 | 514 | 1,531 | 1,572 | 1,744 | 1,413 | 1,568 | 1,794 |
| Thailand | 58 | 96 | 476 | 475 | 450 | 532 | 782 | 1,017 |

*Sources*: Mataloni and Goldberg (1994); Mataloni (1995).

# APPENDIX TABLE 4.A5

## Sales of US MOFAs in developing countries by region and country, 1977–95 (millions of US$)

| | | | | Year | | | | |
|---|---|---|---|---|---|---|---|---|
| Country/Regional | 1977 | 1982 | 1989 | 1991 | 1992 | 1993 | 1994 | 1995 |
| World | 507,019 | 730,235 | 1,019,966 | 1,242,635 | 1,291,649 | 1,275,775 | 1,435,901 | 1,794,089 |
| Developing countries total | 154,260 | 185,215 | 178,933 | 213,107 | 238,979 | 251,068 | 293,981 | 351,924 |
| Latin America and Caribbean | 58,208 | 103,857 | 97,014 | 102,982 | 115,454 | 119,199 | 134,808 | 149,193 |
| Argentina | 3,615 | 5,104 | 4,057 | 5,656 | 7,628 | 8,239 | 11,545 | 12,244 |
| Brazil | 16,630 | 26,045 | 30,588 | 26,988 | 27,820 | 31,852 | 33,232 | 40,005 |
| Chile | 533 | 1,303 | 1,981 | 2,476 | 3,471 | 3,826 | 4,937 | 7,033 |
| Colombia | 1,869 | 4,401 | 3,895 | 4,325 | 4,995 | 5,641 | 6,501 | 7,432 |
| Ecuador | 514 | 803 | 578 | 720 | 611 | 571 | 795 | 1,128 |
| Peru | 867 | 1,828 | 1,122 | 1,409 | 1,279 | 1,399 | 1,632 | 2,413 |
| Venezuela | 4,713 | 7,240 | 2,677 | 3,533 | 4,295 | 4,695 | 5,431 | 6,944 |
| Costa Rica | a | a | 723 | 680 | 1,516 | 1,559 | 1,781 | 1,873 |
| Guatemala | a | a | 672 | 733 | 808 | 732 | 850 | 955 |
| Honduras | a | a | 1,181 | 856 | 970 | 975 | 1,035 | 1,044 |
| Mexico | 5,860 | 11,269 | 16,437 | 24,838 | 30,137 | 32,548 | 39,421 | 35,879 |
| Panama | 1,340 | 3,044 | 1,825 | 1,754 | 1,929 | 1,701 | 1,839 | 1,974 |
| Jamaica | a | | 1,138 | 1,044 | 1,023 | 1,036 | 1,116 | 1,176 |
| Trinidad and Tobago | 1,154 | 2,521 | 652 | 844 | b | b | 760 | 1,112 |
| Africa | 19,023 | 23,596 | 11,576 | 13,513 | 14,113 | 13,533 | 14,866 | 17,651 |
| South Africa | 4,613 | 7,841 | 2,653 | 3,079 | 3,113 | 3,005 | 3,630 | 5,126 |
| Other | 14,410 | 15,755 | 8,923 | 10,434 | 11,000 | 10,528 | 11,236 | 12,525 |
| Middle East | 62,922 | 16,699 | 8,021 | 8,199 | 9,285 | 7,975 | 8,070 | 9,938 |
| Israel | 568 | 589 | 1,042 | 1,586 | 1,724 | 1,970 | 2,351 | 2,850 |
| Other | 62,354 | 16,110 | 6,979 | 6,613 | 7,561 | 6,005 | 5,719 | 7,088 |
| Asia and Pacific | 18,720 | 48,903 | 62,322 | 88,413 | 100,127 | 110,361 | 136,237 | 175,142 |
| China | a | a | 257 | 993 | 1,348 | 1,393 | 3,225 | 5,721 |
| Hong Kong | 4,931 | 7,516 | 16,408 | 19,848 | 21,743 | 24,752 | 29.729 | 38,240 |
| India | 513 | 618 | 323 | 307 | 330 | 450 | 983 | 1,685 |
| Indonesia | 5,468 | 12,543 | 6,120 | 7,891 | 7,963 | 7,768 | 8,229 | 10,018 |
| Korea, Republic of | 219 | 604 | 2,463 | 3,494 | 3,935 | 4,130 | 5,554 | 7,812 |
| Malaysia | 1,148 | 4,319 | 5,419 | 7,415 | 8,786 | 9,526 | 11,579 | 14,098 |
| Philippines | 2,062 | 3,596 | 2,905 | 3,616 | 4,081 | 4,375 | 5,211 | 6,278 |
| Singapore | 2,113 | 14,114 | 15,102 | 28,315 | 34,056 | 38,267 | 46,871 | 60,220 |
| Taiwan | 1,004 | 1,867 | 6,773 | 7,890 | 9,046 | 10,055 | 13,690 | 16,787 |
| Thailand | 916 | 2,590 | 5,456 | 7,465 | 7,496 | 8,171 | 9,627 | 12,506 |

Sources: US Department of Commerce (1981), (1985), (1992), (1994), Mataloni and Fahim-Nader (1996); Mataloni (1997).

Notes:

a Data not available.

b Data in the cell have been suppressed to avoid disclosure of data of individual companies.

# APPENDIX TABLE 4.A6

## Sales of US manufacturing MOFAs in developing countries by region and country, 1977–94 (millions of US$)

| Country/Region | 1977 | 1982 | 1989 | 1991 | 1992 | 1993 | 1994 |
|---|---|---|---|---|---|---|---|
| World | 194,200 | 271,099 | 509,308 | 596,257 | 622,608 | 622,707 | 694,666 |
| Developing countries total | 30,278 | 50,864 | 76,345 | 89,787 | 102,152 | 116,362 | 136,375 |
| Latin America and Caribbean | 24,217 | 39,506 | 48,239 | 53,113 | 61,588 | 68,393 | 76,756 |
| Argentina | 2,466 | 3,223 | 2,717 | 3,454 | 5,014 | 5,097 | 7,166 |
| Brazil | 11,218 | 17,038 | 24,330 | 20,656 | 21,357 | 24,824 | 25,981 |
| Chile | 199 | 383 | 970 | 823 | 987 | 1,336 | 1,789 |
| Colombia | 992 | 1,846 | 1,544 | 1,677 | 2,081 | 2,541 | 2,960 |
| Ecuador | 114 | 181 | 187 | 223 | 243 | 245 | 301 |
| Peru | 310 | 320 | 279 | 216 | 254 | 316 | 411 |
| Venezuela | 2,793 | 4,848 | 1,984 | 2,578 | 3,376 | 3,668 | 3,624 |
| Costa Rica | a | a | 309 | 438 | 542 | 610 | 642 |
| Guatemala | a | a | 197 | 197 | 232 | 257 | 304 |
| Honduras | a | a | 419 | 306 | 371 | 402 | 383 |
| Mexico | 4,679 | 9,438 | 14,246 | 21,523 | 25,731 | 27,232 | 31,008 |
| Panama | 71 | 173 | 184 | 197 | 442 | 257 | 191 |
| Jamaica | a | a | 243 | 139 | 163 | 186 | 219 |
| Trinidad and Tobago | 45 | 74 | 61 | 45 | 49 | 47 | 28 |
| Africa | 2,581 | 4,328 | 2,560 | 2,983 | 3,092 | 3,016 | 3,529 |
| South Africa | 1,940 | 3,398 | 1,357 | 1,488 | 1,516 | 1,447 | 1,875 |
| Other | 641 | 930 | 1,203 | 1,495 | 1,576 | 1,569 | 1,654 |
| Middle East | 295 | 495 | 538 | 963 | 1,100 | 1,357 | 1,779 |
| Israel | 209 | 301 | 485 | 872 | 1,006 | 1,281 | 1,550 |
| Other | 86 | 194 | 53 | 91 | 94 | 76 | 229 |
| Asia and Pacific | 5,125 | 9,933 | 25,008 | 32,728 | 36,372 | 43,596 | 54,311 |
| China | a | a | 121 | 351 | 529 | 750 | 1,890 |
| Hong Kong | 745 | 1,135 | 3,543 | 4,172 | 4,647 | 4,958 | 6,723 |
| India | 467 | 508 | 271 | 266 | 275 | 350 | 740 |
| Indonesia | 262 | 484 | 341 | 445 | 511 | 627 | 1,727 |
| Korea, Republic of | 187 | 414 | 1,518 | 2,122 | 2,368 | 2,459 | 2,990 |
| Malaysia | 445 | 1,618 | 2,681 | 3,858 | 4,833 | 5,350 | 6,680 |
| Philippines | 1,010 | 1,678 | 1,664 | 2,073 | 2,536 | 2,801 | 3,035 |
| Singapore | 882 | 1,877 | 7,579 | 11,425 | 12,317 | 16,984 | 20,290 |
| Taiwan | 782 | 1,496 | 4,879 | 4,916 | 5,262 | 5,911 | 6,080 |
| Thailand | 234 | 521 | 2,132 | 2,836 | 2,811 | 3,104 | 3,838 |

*Sources*: US Department of Commerce (1981), (1985), (1992), (1994), (1995a), (1995b), (1996) and (1997a).

*Note*:
a  Data not available.

# APPENDIX TABLE 4.A7

## US exports shipped to MOFAs by parents, 1977–94
## (millions of US$)

| Country/Region | 1977 | 1982 | 1989 | 1991 | 1992 | 1993 | 1994 |
|---|---|---|---|---|---|---|---|
| World | 35,813 | 52,753 | 86,050 | 95,779 | 100,737 | 106,827 | 125,423 |
| Developing countries total | 6,489 | 12,080 | 16,746 | 20,756 | 23,171 | 27,293 | 33,929[c] |
| Latin America and Caribbean | 3,700 | 6,479 | 9,495 | 12,795 | 14,110 | 15,306 | 18,876 |
| Argentina | 186 | 317 | 243 | 356 | 368 | 526 | 655 |
| Brazil | 724 | 706 | 1,242 | 1,048 | 1,103 | 1,419 | 1,627 |
| Chile | 53 | 63 | – | 160 | 155 | 208 | 244 |
| Colombia | 181 | 239 | 251 | 241 | 301 | 344 | 520 |
| Ecuador | 40 | 84 | 80 | 79 | 52 | 40 | 45 |
| Peru | 83 | 265 | 28 | 39 | 37 | 57 | 83 |
| Venezuela | 1,042 | 1,365 | 326 | 702 | 870 | 1,097 | 873 |
| Costa Rica | [a] | [a] | 77 | 41 | 94 | 94 | 114 |
| Guatemala | [a] | [a] | 51 | 118 | 123 | 32 | 92 |
| Honduras | [a] | [a] | 138 | 110 | 113 | 84 | 90 |
| Mexico | 813 | 2,328 | 6,090 | 8,892 | 10,096 | 10,718 | 13,361 |
| Panama | 79 | 239 | 204 | 277 | 169 | 158 | 207 |
| Jamaica | [a] | 118 | 132 | 162 | 80 | 71 | 161 |
| Trinidad and Tobago | 32 | 75 | 11 | 6 | [b] | [b] | 18 |
| Africa | 648 | 999 | [b] | 336 | 306 | 331 | 429 |
| South Africa | 324 | 524 | 176 | 185 | 167 | 153 | 230 |
| Other | 324 | 475 | – | 151 | 139 | 178 | 199 |
| Middle East | 937 | 632 | 288 | 190 | 187 | 166 | 198 |
| Israel | 28 | 55 | 132 | 34 | 36 | 31 | 68 |
| Other | 909 | 577 | 156 | 156 | 151 | 135 | 130 |
| Asia and Pacific | 1,528 | 4,449 | 6,874 | 7,711 | 8,805 | 10,944 | 14,394 |
| China | [a] | [a] | 35 | 107 | 148 | 309 | 197 |
| Hong Kong | 324 | 936 | 2,130 | 1,893 | 2,746 | 3,352 | 4,971 |
| India | 11 | 12 | 22 | 20 | 18 | 19 | 22 |
| Indonesia | 127 | 387 | 95 | 89 | 163 | 280 | 333 |
| Korea, Republic of | 99 | 172 | 422 | 607 | 631 | 846 | 942 |
| Malaysia | 221 | 969 | 668 | 661 | 744 | 926 | 1,062 |
| Philippines | 132 | 296 | 239 | 146 | 126 | 146 | 254 |
| Singapore | 343 | 955 | 2,067 | 2,391 | 2,485 | 3,136 | 4,603 |
| Taiwan | 166 | 385 | 692 | 902 | 1,053 | 1,190 | 1,211 |
| Thailand | 75 | 296 | 443 | 773 | 658 | 701 | 745 |

*Source*: US Department of Commerce (1981), (1985), (1992), (1994), (1995), (1996) and (1997a).

*Notes*:
a  Data not available.
b  Data in the cell have been suppressed to avoid disclosure of data of individual companies.
c  Includes international. Minimum estimates, excluding international, were 16,922 for 1989 and 33,897 for 1994.

# APPENDIX TABLE 4.A8

## US manufacturing exports shipped to MOFAs by parents, 1977–94 (millions of US$)

| Country/Region | Year | | | | | | |
|---|---|---|---|---|---|---|---|
| | 1977 | 1982 | 1989 | 1991 | 1992 | 1993 | 1994 |
| World | 25,145 | 34,748 | 57,707 | 62,915 | 65,272 | 66,051 | 74,578 |
| Developing countries total | 3,717 | 7,305 | 11,617 | 14,651 | 16,121 | 17,625 | 20,431 |
| Latin America and Caribbean | 2,644 | 4,379 | 7,613 | 10,241 | 11,700 | 12,481 | 14,644 |
| Argentina | 156 | 250 | 190 | 181 | 181 | 233 | 241 |
| Brazil | 618 | 588 | 1,185 | 979 | 1,064 | 1,064 | 1,310 |
| Chile | 20 | 20 | b | 43 | 42 | 54 | 90 |
| Colombia | 131 | 121 | 121 | 116 | 162 | 173 | 258 |
| Ecuador | b | b | b | b | 26 | 19 | 20 |
| Peru | 23 | b | b | 20 | 21 | 26 | 27 |
| Venezuela | 749 | 860 | 201 | 502 | 678 | 944 | 606 |
| Costa Rica | a | a | 30 | 30 | 30 | 32 | 98 |
| Guatemala | a | a | 11 | 13 | 15 | 15 | 13 |
| Honduras | a | a | 48 | 13 | 20 | 20 | 10 |
| Mexico | 725 | 2,109 | 5,667 | 8,215 | 9,335 | 9,775 | 11,831 |
| Panama | 17 | 36 | b | b | 20 | 25 | 28 |
| Jamaica | a | 95 | b | 19 | 30 | 27 | 26 |
| Trinidad and Tobago | b | 17 | b | b | 4 | 3 | b |
| Africa | 235 | 387 | 225 | 158 | 147 | 129 | 228[c] |
| South Africa | 140 | 287 | b | 123 | 4 | 97 | 112 |
| Other | 95 | 100 | – | 35 | 143 | 32 | – |
| Middle East | 43 | 54 | 16 | 28 | 23 | 26 | – |
| Israel | b | 33 | b | b | b | 9 | 34 |
| Other | – | 21 | – | – | – | 17 | – |
| Asia and Pacific | 935 | 2,771 | 3,763 | 4,224 | 4,251 | 4,996 | 5,559 |
| China | a | a | 11 | 15 | 45 | – | 106 |
| Hong Kong | 126 | 219 | 416 | 606 | 581 | 574 | 642 |
| India | 8 | 11 | 4 | 7 | 6 | 6 | 13 |
| Indonesia | 29 | 106 | b | 38 | 26 | 27 | 69 |
| Korea, Republic of | 89 | 147 | 239 | 338 | 406 | 453 | 460 |
| Malaysia | 200 | 888 | 570 | 533 | 599 | 736 | 759 |
| Philippines | 65 | 244 | 194 | 109 | 82 | 98 | 176 |
| Singapore | 218 | 561 | b | 1,490 | 1,530 | 1,886 | 2,080 |
| Taiwan | 142 | 348 | 493 | 473 | 513 | 620 | 660 |
| Thailand | 49 | 231 | 314 | 609 | 457 | 505 | 583 |

*Sources*: US Department of Commerce (1981), (1985), (1992), (1994), (1995), (1996) and (1997a).

*Notes*:
a Data not available.
b Data in the cell have been suppressed to avoid disclosure of data of individual companies.
c Includes Middle East.

# APPENDIX TABLE 4.A9

## Manufacturing gross product of MOFAs, 1994
## (millions of US$)

| Country/Region | Total mfg | Food and kindred products | Chemicals and allied products | Primary and fabricated metals | Industrial machinery and equip | Electronic and other electrical equipt | Trans equipt | Other mfg |
|---|---|---|---|---|---|---|---|---|
| World | 205,208 | 24,750 | 40,970 | 8,051 | 27,490 | 19,866 | 35,886 | 48,195 |
| Developed countries | 163,875 | 19,003 | 32,532 | 6,528 | 23,217 | 13,505 | 29,222 | 39,868 |
| Developing countries | 41,333 | 5,747 | 8,438 | 1,523 | 4,273 | 6,361 | 6,664 | 8,327 |
| Latin America and Caribbean | 27,615 | 4,732 | 6,348 | 1,117 | 1,135 | 1,827 | 6,100 | 6,357 |
| Argentina | 2,311 | 992 | 650 | 70 | 48 | 47 | 29 | 476 |
| Brazil | 13,450 | 1,336 | 2,805 | 555 | 821 | 386 | 3,956 | 3,591 |
| Chile | 657 | 78 | 143 | 212 | * | 8 | 83 | 133 |
| Colombia | 959 | 200 | 338 | 30 | −1 | b | b | 227 |
| Ecuador | 80 | 37 | 24 | 4 | 0 | * | 0 | 14 |
| Peru | 105 | 25 | 47 | 2 | 0 | 3 | 0 | 29 |
| Venezuela | 1,005 | 215 | 215 | 41 | 19 | b | b | 303 |
| Costa Rica | 187 | 58 | 41 | 14 | 0 | b | 0 | b |
| Guatemala | 95 | 27 | 21 | 6 | 0 | 0 | 0 | 41 |
| Honduras | 146 | b | 3 | 2 | 0 | 0 | 0 | b |
| Mexico | 8,104 | 1,440 | 1,922 | 178 | 329 | 1,304 | 1,701 | 1,232 |
| Panama | 70 | 13 | 20 | 4 | 0 | 0 | 0 | 34 |
| Jamaica | 95 | 0 | 46 | 0 | 0 | 0 | 0 | 49 |
| Africa | 1,065 | 302 | 369 | 105 | 39 | 12 | 21 | 216 |
| South Africa | 548 | 88 | 251 | 33 | 36 | 5 | 21 | 116 |
| Other | 517 | 214 | 118 | 72 | 3 | 7 | 0 | 100 |
| Middle East | 706 | 90 | 45 | 8 | 6 | 475 | 2 | 81 |
| Israel | 656 | b | 29 | 3 | 1 | 470 | 0 | b |
| Other | 50 | b | 16 | 6 | 5 | 5 | 0 | b |
| Asia and Pacific | 11,947 | 623[a] | 1,,676 | 293 | 3,093 · | 4,047 | 541[a] | 1,673[a] |
| China | 488 | 29 | 108 | 33 | 86 | 172 | 0 | 62 |
| Hong Kong | 1,316 | 19 | 87 | 76 | 125 | 585 | 21 | 404 |
| India | 217 | b | 82 | b | 44 | −2 | * | 80 |
| Indonesia | 390 | 16 | 254 | 8 | b | 19 | 10 | b |
| Korea, Republic of | 762 | 85 | 131 | 8 | 52 | 231 | 21 | 234 |
| Malaysia | 1,632 | 20 | 80 | b | 219 | 1,039 | 0 | b |
| Philippines | 969 | 240 | 353 | b | 5 | 217 | 0 | b |
| Singapore | 3,727 | 58 | 138 | 38 | 2,241 | 1,008 | 115 | 131 |
| Taiwan | 1,568 | 83 | 237 | b | 78 | 604 | b | 174 |
| Thailand | 782 | 62 | 164 | 75 | b | 167 | 0 | b |

*Source*: US Department of Commerce (1997a).

*Notes*:

a Figures for New Zealand were suppressed in the source – gross product of $175 million in foods, $12 million in transport equipment and $81 in other mfg were estimated by distributing the $268 million not reported by industry in proportion to the mid-point of employment brackets from US Dept. of Commerce (1997), Table III.G3.
b Data in the cell have been suppressed to avoid disclosure of data of individual companies.
* Less than $500,000.

128

# 5

# GERMAN MNEs

## Consequences of eastward drive for developing countries

*Jamuna P. Agarwal*

## Introduction

Foreign direct investment (FDI) of German enterprises has registered a very strong growth since the beginning of this decade. But this growth has been rather uneven among different host regions. The share of developed countries in German outward FDI has declined, whereas the shares of Central and Eastern Europe (CEE) as well as of developing countries have increased. Between the latter two groups of countries, CEE has recorded a considerably higher growth of German FDI. This has been a source of concern to developing countries. They are apprehensive that some of German equity capital may be diverted from their economies to Central and Eastern European countries (CEECs), especially because some of the locational advantages such as low labour costs of these two groups of countries are similar. In addition, many CEECs have secured trade preferences for their exports to German market through 'Europe Agreements', and are geographically and culturally nearer to German investors than most of developing countries. This chapter attempts to dispel the concern about investment diversion from developing countries. For this purpose, at first the recent growth and regional structure of German FDI are analysed. This is followed by a discussion of the main causes of the relatively high involvement of German firms in CEE. Finally, we argue that German FDI in CEE is not likely to occur at the cost of developing countries.

## High growth of FDI outflows

Table 5.1 highlights the growth of German FDI. As compared to the second half of the 1980s, German MNEs raised their average outward investments until 1995 by a factor of 2.6. This has lifted their share in global FDI outflows from 7.9 per cent to 9.2 per cent. Germany ranks now after the US, UK and Japan as the most important global investor. In contrast to Germany,

*Table 5.1* Relative strength of German FDI outflows in the world, 1984–95

|  | Annual averages | |
|---|---|---|
|  | 1984–9 | 1990–5 |
| German outward FDI (mill. $) | 9,599 | 25,131 |
| German share in global FDI outflows | 7.9 | 9.2 |
| German share in total FDI outflows from developed economies | 8.4 | 10.4 |
| German share in total FDI outflows from the European Union | 15.3 | 19.4 |
| Share of developed economies in global FDI outflows | 93.7 | 88.2 |
| Share of European Union in global FDI outflows | 51.5 | 47.2 |
| Share of the US in global FDI outflows | 13.9 | 21.7 |
| Share of Japan in global FDI outflows | 17.1 | 11.8 |

*Source:* UNCTAD (1996a).

shares of the European Union (EU), Japan, and developed countries taken together in total world FDI outflows declined during the corresponding period. The major exception is the US whose FDI outflows have risen faster (UNCTAD, 1996a: 233–7).

The high growth of FDI outflows is argued sometimes to be a result of declining locational competitiveness of the German economy, especially in view of a very low level of FDI inflows (Table 5.2). The average negative yearly balance of FDI flows in Germany has gone up from $8 billion (1984–9) to $19 billion (1990–5). A negative balance of FDI flows is, however, not a reliable indicator of locational competitiveness of a country. In the case of Japan, for example, its locational competitiveness was until recently beyond doubt, but it always had a high negative balance in FDI flows. Nevertheless, wage costs per hour in Germany are high, and have increased recently more than in many competing countries. This is also a result of a long-run trend of rising real exchange rate of the German currency. This may have encouraged local firms to invest abroad, and discouraged foreign firms from investing in Germany. Other important reasons for meagre FDI inflows into Germany are a relatively low market capitalization indicated by the ratio of equity capital of German enterprises listed at the stock exchange to GDP, and high cross-holdings among them (Deutsche Bundesbank, 1997: 28–9). This makes acquisition of German companies by foreign investors rather difficult. In the case of developed countries, mergers and acquisitions are the most popular mode of FDI. Nine tenths of US equity capital outflows in 1995 were implemented through mergers and acquisitions in the host countries (UNCTAD, 1996a: 11). Under such circumstances, it is obvious that FDI inflows in Germany are likely to be less than in countries having a greater scope for mergers and acquisitions by foreign investors (Klodt and Maurer, 1996: 27–8).

130

*Table 5.2* Net balance of FDI flows in Germany, 1984–95 (mill. $)

| Year | Inflows | Outflows | Balance |
|------|---------|----------|---------|
| 1984–9 (annual average) | 1,833 | 9,599 | −7,766 |
| 1990 | 2,689 | 24,214 | −21,525 |
| 1991 | 4,071 | 23,723 | −19,652 |
| 1992 | 2,370 | 19,698 | −17,328 |
| 1993 | 277 | 13,176 | −12,899 |
| 1994 | −2,993 | 14,653 | −17,646 |
| 1995 | 8,996 | 35,302 | −26,306 |

*Source*: UNCTAD (1996a).

## Regional focus on Central and Eastern Europe

More interesting from the point of view of this chapter is the development of the regional structure of German outward FDI. The overwhelming portion of German FDI continues to go to other developed countries – mostly to the EU members. But their share has been declining (Table 5.3). Only to Japan is there some evidence that German FDI has increased in the recent past. The governments of both Germany and Japan are striving to raise the level of capital involvement by German MNEs in the Japanese economy. Germany wants to strengthen its market access through FDI and Japan wants to lend more credibility to its liberal stance for incoming FDI.

The highest growth of German FDI has been recorded since 1990 in CEE. Considering the initial low level, some of this growth is not very dramatic.

*Table 5.3* Regional distribution of German outward FDI, 1990–4

| Country/Region | Stock 1990 (mill. DM) | Share % | Outflows (annual averages) 1991–2 (mill. DM) | Share % | 1993–4 (mill. DM) | Share % |
|----------------|----------------------|---------|----------------|---------|----------------|---------|
| Total | 226,462 | 100 | 30,701 | 100 | 30,197 | 100 |
| Developed countries | 202,724 | 89.5 | 25,953 | 84.5 | 23,017 | 76.2 |
| European Union (15) | 121,894 | 53.8 | 18,049 | 58.8 | 13,927 | 46.1 |
| USA | 53,693 | 23.7 | 6,419 | 20.9 | 6,708 | 22.2 |
| Japan | 4,562 | 2.0 | 712 | 2.3 | 900 | 3.0 |
| Central and Eastern Europe | 762 | 0.3 | 1,210 | 4.0 | 2,454 | 8.1 |
| Developing countries | 22,976 | 10.2 | 3,537 | 11.5 | 4,726 | 15.7 |
| Africa | 3,572 | 1.6 | 49 | 0.2 | 126 | 0.4 |
| Asia | 5,757 | 2.6 | 1,246 | 4.1 | 796 | 2.6 |
| Latin America | 13,647 | 6.0 | 2,242 | 7.3 | 3,805 | 12.6 |

*Source*: Deutsche Bundesbank (1995 and 1996).

However, the share of CEE in German FDI outflows jumped to 13 per cent in 1994 from less than 1 per cent in 1990. This is higher than the share of CEE in FDI of the EU, OECD or any of their member countries except Austria (Table 5.4). Austrian MNEs are more active in CEE than German MNEs. Reasons given below for a relatively high engagement of German investors are often relevant for the Austrian case, too. Germany is centrally located in Europe. Geographical distances to many of the CEECs compared with those of other bigger international investors are shorter. Germany has common borders with some of the CEECs (Poland and the Czech Republic), which generally exercise a strong positive influence on FDI flows between neighbouring countries (Agarwal, 1994). Culturally too, most CEECs are not very alien to German investors. Moreover, they share long-standing trade relations. The eastern part of Germany was a member of the former trade block of communist countries (Council of Mutual Economic Assistance, CMEA), and had a large part of its trade with CEECs. West Germany was the most important western economic partner of the CMEA. Since its collapse in 1989, Germany has evolved into the dominant trader, investor, aid donor and creditor of CEE (Hoey, 1996: 5). Thus, Germany is not only a natural trade, but also a 'natural' investment, partner of these countries.

Trade and investment flows between Germany and CEE may have been facilitated by the envisaged political and economic integration of most of the CEECs in the EU. The 'Europe Agreements' signed with Bulgaria, the Czech Republic, Estonia, Hungary, Latvia, Lithuania, Poland, Romania, Slovakia and Slovenia allow them to become full members of the EU in future depending on the progress of their economic and political convergence. In the meantime, these agreements provide for preferential trading, investment promotion, technical and financial assistance, scientific, industrial and monetary cooperation (Hiemenz et al., 1994). Trade preferences and economic

*Table 5.4* Share of CEECs in FDI outflows of OECD members,[a] 1994 (percentage)

| Country | 1994 | Country | 1994 |
|---|---|---|---|
| Austria | 42.5 | Japan | 0.4 |
| Belgium– | | Netherlands | 2.2 |
| Luxembourg | 2.5 | Norway | 1.7 |
| Denmark | 1.4 | Spain | 0.5 |
| Finland | 0.9 | Sweden | 1.2 |
| France | 2.3 | Switzerland | 2.2 |
| Germany | 13.0 | Turkey | 4.9 |
| Iceland | 8.5 | UK | 1.2 |
| Italy | 2.8 | USA | 1.5 |

*Source*: OECD (1996).

*Note*: a Data for the members not included here are not available.

integration are known to promote FDI (Agarwal, 1994; Barell *et al.*, 1996). Classical cases of trade preferences are found among the former imperial powers like the UK and its colonies. The 'imperial preferences' led to high FDI inflows in the colonies from the UK. The preferential agreements of the EU with Mediterranean countries resulted in many cases in more involvement of EU firms (Joekes, 1982; Pomfret, 1986). The same applies for the extension of the EU to Spain and Portugal. In so far as Germany is involved as a significant donor and trading partner in the implementation of the 'Europe Agreements', they may have helped German firms to invest in CEE.

The transformation of CEECs from centrally planned to market oriented economies awakened prospects of economic growth in this region. Some of them have already been able to record respectable rates of economic growth. Ongoing progress in macroeconomic reforms has improved prospects of future growth of GDP in many of the CEECs. Growth is a significant determinant of inflows of FDI. In the case of CEECs, FDI inflows have often been more volatile than justified by changes in rates of economic growth and their future prospects. This is ascribed to FDI inflows based on privatization of state-owned enterprises in these countries. Privatization-induced FDI flows are lumpy and can lead to heavy swings, especially in relatively small economies. Therefore, the link between economic growth and FDI inflows applies to non-privatization FDI flowing into CEECs (UNCTAD, 1996a: 66–7). Privatisation of state enterprises has attracted a substantial portion of FDI in CEECs. It accounted for as much as 60 per cent of their inflows during 1991 to 1993 (UNCTAD, 1995a: 18). A year later, the privatization of state enterprises contributed only 18 per cent of FDI inflows in this region (UNCTAD, 1996a: 65).

Growth and privatization are likely to have stimulated FDI inflows in CEE from all sources including Germany. However, it is possible that German firms have had comparative advantages in exploiting these locational factors in CEE owing to relatively lower transaction costs emanating from the first two reasons mentioned earlier for high German FDI in this region. Low geographical and cultural distances as well as intensive trade relations involving established distribution and servicing networks weigh heavily in dampening transaction costs of investors. Austria has common borders with four CEECs (the Czech Republic, Hungary, Slovakia and Slovenia) as compared to Germany with only two, and is geographically nearer than Germany to other CEECs. Culturally, both Germany and Austria share same affinities with these countries. So it is not surprising that CEE's share in FDI outflows of Austria is a multiple of their share of German FDI outflows (Table 5.4). Moreover, until 1988 Austria was an insignificant international investor accounting only for 0.2 per cent of average global outflows between 1983 and 1988 (UNCTAD, 1995a: 397). Since the acceleration of its FDI into the CEECs, ignited by their economic transformation from planned into market economies, Austria nearly quadrupled its average share in world FDI outflows between 1989 and 1994 (ibid.). FDI into CEE amounted to 30 per cent of all Austrian FDI outflows during the same period (OECD, 1996). This means that (1) CEE has contributed significantly in lifting Austrian importance as an international investor, and (2) Austrian FDI in CEE is new, occurring on its own merits and not at the cost of the developing or any other group of countries. This is easier to see in the Austrian than German case because Austrian FDI involvement in developing countries prior to the take off of its FDI in CEE was minuscule. Therefore, it is hardly thinkable that Austrian FDI has been diverted from them to CEECs.

## Growth of FDI in CEE – a creation or diversion of investment?

Like Austria, CEE's share in German FDI outflows and German share of total world FDI outflows have increased simultaneously (Tables 5.1 and 5.3). If the increase in flows to CEE since 1989 are excluded from total German FDI outflows, more than one half of the rise in German share of global outflows would be mitigated. This means that the market orientation of CEECs has motivated German investors to create additional equity capital for investment in this region, and not to divert from developing countries. The assumption that German FDI has not been diverted from developing countries to CEE seems to be rather realistic also because the former's share in total German FDI outflows has considerably increased during the relevant period rather than decreasing (Table 5.3). The latter should have occurred, *ceteris paribus*, in the case of investment diversion. As an approximation, no diversion hypothesis appears to be realistic also on the ground that nearly nine tenths of German FDI outflows have gone to only three countries – the Czech Republic, Poland and Hungary (Table 5.5). Of these the former two have common borders with Germany, and Hungary is the most advanced CEE country in terms

*Table 5.5* Distribution of German FDI in Central and Eastern Europe, 1989–95

| Country | FDI (%) |
| --- | --- |
| Czech Republic[a] | 33.9 |
| Hungary | 39.8 |
| Poland | 15.6 |
| Russian Federation[b] | 2.9 |
| Other countries | 7.8 |
| Total | 12,750 mill. DM |

*Source*: Transferstatistik of the German Ministry for Economic Affairs quoted in Beifuß (1996).
*Notes*:
a 1989–92 Czechoslovakia.
b 1989–91 Former Soviet Union.

of economic reforms and is already a member of OECD. As discussed elsewhere (Agarwal, 1994), dismantling tariff and non-tariff barriers to trade and allowing freer movement of capital across borders boost FDI flows between neighbouring countries. This was observed between Portugal and Spain, and between Spain and France after the former two joined the EU. The same can be maintained about Mexico and NAFTA, and about the members of MERCOSUR (IADB-IRELA, 1996). Geographical and cultural vicinity tend to reduce transaction costs which play a crucial role in investment decisions specially of small and medium size enterprises (SMEs). For example, many SMEs from Bavaria in South Germany are investing in the Czech Republic (DIHT, 1996: 13–14), which would have not internationalized their production in the absence of an opening of the Czech border.

It is interesting to observe that the expansion of German firms in CEE does not appear to be at the expense of domestic investment either. The ratios of FDI outflows to gross fixed capital formation in Germany have been declining rather than rising since these flows started to grow rapidly in this region (Table 5.6). The counterfactual is, of course, difficult, if not impossible, to prove. However, it appears realistic that foreign expansion of German firms as a result of both the economic and political transformation of its neighbours in CEE and the global trend towards internationalization of production and markets has created more equity capital rather than diverting it from domestic locations or developing countries.

The decline in German FDI outflows to EU members is related to the Single Market programme. Intra-union German FDI outflows were boosted by the Single Market as German firms invested in other European economies in search of scale economies and to secure a greater share in services markets (banking, insurance, transport and communication). Services were more protected against foreign competition in the member countries than goods market before the Single Market programme (Nunnenkamp et al., 1994: 11–17). After the completion of the Single Market programme in 1992, the related equity capital outflows subsided. Market structures of the developed EU

*Table 5.6* Ratio of FDI outflows to gross fixed capital formation in Germany, 1984–94

| Year | FDI ratio (%) |
| --- | --- |
| 1984-9 | 5.2 |
| 1990 | 7.7 |
| 1991 | 7.0 |
| 1992 | 5.2 |
| 1993 | 3.9 |
| 1994 | 4.2 |

*Source*: UNCTAD (1996a).

economies and CEECs are still so very different that it seems to be erroneous to surmise about a diversion of German investment from the former to the latter.

## How unlikely is investment diversion in future?

The answer to the question of future investment diversion relies on the sectoral hypothesis (Agarwal, 1994). It distinguishes between location-bound and location-mobile FDI. The former type of FDI is attracted by the availability of natural resources or the need for immediate market presence of the investor in a host country. Its susceptibility to diversion from one location to another in response to changes in their locational conditions is very low, if not non-existent. The latter type of FDI, which is considered as internationally mobile between different locations, is a function of relative production costs in host countries. It is commonly termed as efficiency-oriented FDI as compared to the former, which is also called natural resource or market-oriented FDI. Location-mobile FDI is divertable by shifts in relative costs and conditions of competing countries. In this type of FDI, a firm is under pressure to raise the cost-efficiency of its product to sustain international competition, and will tend to divert its existing but more often future production from a given location to another country becoming more cost competitive. Location-mobile FDI is found in labour intensive branches or parts of value added chains in manufacturing sector. The recent technological revolution allows firms selectively to separate these parts and locate their production in different countries according to their comparative costs. Similarly, the revolution in data processing and communication has made production of some of the services (e.g. book-keeping) internationally mobile. So FDI in such services is likely to be location-mobile and subject to diversion. Otherwise, FDI in the service sector is traditionally considered as market-seeking and not prone to diversion from one to another country. Since location-mobile FDI constitutes worldwide only a small percentage of total FDI,[1] investment diversion which may occur in this case is considered as negligible when compared to the rest of FDI which is location-bound and non-divertable (Agarwal, 1996a).

A wide-ranging empirical verification of the sectoral hypothesis is impracticable due to lack of data on regional flows of location-mobile FDI, particularly those which may be inspired by the recent technological revolution. It is not known for Germany or any other country how much of FDI by different industries is undertaken to produce labour intensive parts or processes of their value added chains in developing countries or CEE. Automobile and electronic industries are observed to be striving for relocation of some of their labour intensive production in countries having low labour costs. But separate data on such FDI are not available. Similar problems exist in the service sector. The US is the only country publishing data on FDI in 'computer processing and data preparation services'. How much of this is motivated to utilize low costs of production in host countries is again not known. Regardless of this fact 'computer processing and data preparation services' accounted for only 0.4 per cent of outward FDI stock of all services and 0.2 per cent of total stock of US FDI in 1995 (US Department of Commerce, 1996: 127). This shows how narrow the scope for an eventual investment diversion in this subsector is.

In view of the data constraints, an analysis of German FDI diversion from developing countries to CEE has to concentrate on traditional labour intensive industries of textile, leather and clothing. In these industries, German equity capital can be considered as mobile between different host countries in response to their relative factor prices. The liberalization of investment regimes in CEECs together with their relatively low unit wage costs have widened locational choices for German firms in these industries. To what extent these choices have been availed of by them is not quantifiable. Textile, leather and clothing contribute less than 1 per cent to German stock of all FDI (Table 5.7). This share has remained constant between 1989 and 1994. Moreover, the regional distribution of FDI in these industries is not discernible from the published data. So it is not determinable if there have been any changes in relative shares of developing countries *vis-à-vis* CEECs.[2] But a less than 1 per cent share of these industries in total stock of German FDI indicates a very limited scope for investment diversion from developing countries to CEECs. The data on inward FDI in CEECs also underline the unimportance of textile, leather and clothing. In Estonia and Hungary, they attracted less than 3 per cent and in Poland about 8 per cent of inward FDI from all sources including Germany (Table 5.8). In developing countries too, these industries account for very low shares of inward FDI, mostly below 3 per cent. Only in Indonesia, Malaysia and El Salvador have textile, leather and clothing hosted higher shares of inward FDI stocks (Table 5.9). Furthermore, shares of labour intensive industries in EU and US FDI in Asian developing countries have increased rather than decreasing (EC-UNCTAD, 1996: 40).[3] This goes against the investment diversion hypothesis. This is specially noteworthy in the case of Asian countries which have been more concerned about the investment diversion issue than any other regional group of developing economies.

Nearly all German FDI (99 per cent) is location bound (Table 5.7), and unlikely to be diverted from developing countries to CEECs. This applies to FDI in primary sector, manufacturing excluding textile, leather and clothing, and services.

In the primary sector German FDI has declined in 1994 as compared to 1989 in terms of its share in total outward FDI stock (Table 5.7). Competition between developing countries and CEECs in the primary sector is not visible. The mining sector of the latter group of countries has attracted so far only 0.1 per cent of all German FDI as compared to 2.8 per cent in developing countries (Table 5.10). FDI in the primary sector is highly locational, and cannot be diverted from one country to another unless the latter has the same or better quality of a particular natural resource to offer under competitive costs. Oil is a case in point. Russia and some other CIS members could compete with oil exporting developing countries in attracting FDI. But there is no significant evidence of German capital flowing into this sector of CEE or being diverted from developing countries. FDI of German oil MNEs on the whole does not appear to have changed significantly since 1989.

*Table 5.7* Sectoral distribution of German stock of FDI abroad, 1989 and 1994 (percentage)

| Sector | 1989 | 1994 |
|---|---|---|
| **Total (mill. DM)** | 205,563 | 329,757 |
| **Primary sector** | 1.91 | 1.21 |
| Agriculture | 0.07 | 0.22 |
| Mining and quarrying[a] | 0.70 | 0.16 |
| Oil | 1.15 | 0.82 |
| **Manufacturing sector** | 55.51 | 32.92 |
| Food, beverages and tobacco | n.a. | 0.87 |
| Textiles, leather and clothing | 0.85 | 0.84 |
| Paper, printing and publishing[b] | 0.96 | 0.53 |
| Chemical products[c] | 19.55 | 11.23 |
| Coal and petroleum products[d] | 0.34 | 0.02 |
| Non-metallic products[e] | 1.40 | 0.97 |
| Metal products | 3.27 | 1.98 |
| Mechanical equipment | 6.68 | 3.06 |
| Electric and electronic equipment[f] | 10.56 | n.a. |
| Motor vehicles[g] | 8.13 | 5.86 |
| Other transport equipment | n.a. | 0.23 |
| Other manufacturing | 2.21 | 7.27 |
| **Services sector** | 38.09 | 65.81 |
| Construction | 0.56 | 0.36 |
| Wholesale and retail trade | 3.75 | 13.71 |
| Transport and storage[h] | 0.97 | 0.85 |
| Finance, insurance and business services | 15.50 | 23.78 |
| Other services[i] | 17.32 | 27.10 |
| **Unallocated** | 4.48 | 0.04 |

*Source*: OECD (1996).

*Notes*:
n.a. = Not available.
a Including manufacture of coke oven products except quarrying;
b Except publishing;
c Including processing of nuclear fuel;
d Except manufacture of coke oven products, processing of nuclear fuel;
e Except manufacture of other non-metallic mineral products;
f Including office equipment;
g Including other vehicles;
h Including communication;
i Including hotels, restaurants, publishing, real estate and holding.

*Table 5.8* Sectoral structures of foreign direct investment (stock) in selected CEE countries, 1994 (per cent)

| Sector | Bulgaria[c] | Estonia | Hungary | Poland[a1] | Romania[a2] | Czech Rep.[e] | Slovakia[a3] |
|---|---|---|---|---|---|---|---|
| **Primary sector** | 1.6 | 2.0 | 1.8 | – | 7.9 | – | 1.1 |
| Agriculture | 0.1 | 1.8[g] | 0.5 | – | 4.1 | – | 0.1[b] |
| Mining and quarrying | 1.5 | 0.2 | 1.3 | – | 3.8 | – | 1.0 |
| **Manufacturing sector** | 38.7 | 35.6 | 48.4 | 75.4 | 39.1[f] | 66.6 | 50.0 |
| Food | – | 13.9[h] | 17.3 | 14.3 | 9.4 | 9.8 | – |
| Textiles, leather and wearing apparel | – | 2.6 | 2.0 | 8.1 | – | – | – |
| Wood, paper, printing | – | 3.6 | 2.4 | 6.9 | – | – | – |
| Chemicals | – | 0.8 | 4.8 | 8.3 | – | 6.0 | – |
| Non-metallic products | – | 8.2 | 4.4 | 0.1 | – | – | – |
| Basic metals and metal products | – | 0.8 | 3.2 | 6.2 | – | – | – |
| Machinery and equipment, n.e.c. | – | 1.1 | 14.4 | 1.4 | – | – | – |
| Electrical equipment | – | 2.0 | – | 5.7 | 3.8 | – | – |
| Transport equipment | – | 0.1 | – | 16.0 | – | 20.3 | – |
| **Services sector** | 57.7 | 62.4 | 49.8 | – | 53.0 | 27.9 | 48.9 |
| Construction | 19.8 | 0.8 | 2.6 | 2.7 | 4.8 | 12.8 | 3.1 |
| Trade | 13.9 | 31.7 | 8.3 | 11.2 | 21.2 | 5.0 | 29.2 |
| Hotels and restaurants | – | 2.6 | 2.6 | 1.1 | 8.8 | – | 0.2 |
| Transport | 21.7[d] | 15.1 | 22.2[d] | 4.1[d] | 8.0 | – | 0.6 |
| Financial intermediation | – | 5.6 | 8.8 | 0.5 | 3.0 | 10.1 | 7.9 |

*Sources*: UN (1992), UN (various issues), OECD (1996).

*Notes*:
a1 1993; a2 between 20/3/1990 and 20/10/1993 (tentative); a3 1992;
b including fishing;
c undistribted: 2 per cent;
d including telecommunications;
e undistributed: 17.6 per cent;
f 4.1 per cent light industries;
g including fishing;
h including tobacco.

Excluding labour intensive industries, German FDI in the manufacturing sector of host countries is attracted by their market size and growth – advantages related to the proximity of customers, and the avoidance of discriminatory government procurement policies, trade barriers and high transport costs. These factors are not affected by the change in locational conditions of CEE *vis-à-vis* developing countries. Therefore, German FDI in most of the manufacturing sector of developing countries is considered as location-bound and is not expected to be affected by the growing interest of German MNEs in CEE. This certainly applies to countries like China, India, Indonesia, Argentina, Brazil, Mexico and South Africa, which have large domestic markets. The size of domestic market has generally been the most important determinant of FDI inflows (Agarwal, 1980; UNCTC, 1992). High growth

*Table 5.9* Share of textiles, leather and clothing in inward FDI stock (percentages)

| Country | Year | Latest available % |
|---|---|---|
| Indonesia | 1993 | 7.1 |
| Malaysia | 1993 | 4.8 |
| Philippines | 1995 | 2.2 |
| Korea, Republic | 1994 | 2.5 |
| Argentina | 1992 | 1.5 |
| Bolivia | 1990 | 2.0 |
| Brazil | 1992 | 1.7 |
| Chile | 1990 | 0.8 |
| Colombia | 1992 | 0.9 |
| Dominican Republic | 1990 | 0.6 |
| El Salvador | 1990 | 12.4 |
| Mexico | 1994 | 1.2 |
| Peru | 1990 | 2.6 |
| Venezuela | 1995 | 0.7 |

*Sources*: UNCTAD (1994b); Central Bank of the Philippines (unpublished); EC-UNCTAD (1996); OECD (1996); IADB-IRELA (1996).

rates and increased income levels in many smaller countries such as Hong Kong, Republic of Korea, Malaysia, Singapore, Thailand, Taiwan and Chile have increased the size of their domestic markets and made them very attractive for foreign investors. The above mentioned countries account for more than four fifths of the German stock of FDI (Deutsche Bundesbank, 1996). Thus, there is hardly any scope for FDI diversion from developing countries to CEE in non-labour intensive manufacturing FDI of German MNEs. Manufacturing is the most important sector for CEECs, as well as developing countries, in attracting German FDI. It hosts 70 per cent of the total stock of German FDI in the former and 64 per cent in the latter group of countries (Table 5.10).

The service sector in German FDI data includes construction, trade, transport, storage, communication, finance, insurance, business services, hotels, restaurants, real estate, etc.[4] In these cases the production and consumption of services have to occur generally within the host country or region (Lipsey and Weiss, 1981). Therefore, German enterprises seeking market shares of these services in CEECs or developing countries will have to invest directly in the respective market places. They cannot serve one market from a location in another country unless they are geographically very close to each other and have no trade restrictions on services, which is not the case between CEECs and developing countries. Mostly they are geographically far apart, and trade between them is more restricted than their trade with developed countries owing to various preferential arrangements (Europe Agreements,

*Table 5.10*  Sectoral distribution of German FDI stock in Central and Eastern Europe (CEE), developed and developing countries, 1989 and 1994 (mill. DM and percentages)[a]

| Sector | CEE | | Developed countries | | Developing countries | |
|---|---|---|---|---|---|---|
| | 1989 | 1994 | 1989 | 1994 | 1989 | 1994 |
| Total FDI (mill. DM) | 477 | 8,960 | 182,722 | 300,664 | 20,762 | 38,633 |
| Mining | n.a. | 0.1 | 1.6 | 0.8 | 4.1 | 2.8 |
| Manufacturing | 84.3 | 70.3 | 41.1 | 36.2 | 68.3 | 63.9 |
| Chemical industry | 3.8 | 7.2 | 16.0 | 13.3 | 18.0 | 16.3 |
| Machinery | 11.7 | 2.7 | 3.8 | 3.6 | 6.4 | 6.1 |
| Automobile | n.a. | 23.6 | 4.5 | 5.2 | 21.0 | 20.4 |
| Electricals | 2.9 | 6.6 | 6.8 | 5.5 | 12.8 | 11.9 |
| Other industries | 65.9 | 30.2 | 10.0 | 8.6 | 10.1 | 9.2 |
| Trade | 3.6 | 14.6 | 23.4 | 17.0 | 7.7 | 8.4 |
| Banking | n.a. | 6.1 | 6.7 | 9.0 | 7.8 | 11.1 |
| Finance excl. banks | n.a. | 0.3 | 7.4 | 14.0 | 6.2 | 3.7 |
| Insurance | n.a. | 1.1 | 4.2 | 6.2 | 0.9 | 1.5 |
| Unallocated | n.a. | 7.5 | 15.6 | 16.8 | 5.0 | 8.7 |

*Source:* Deutsche Bundesbank (1994 and 1996).

*Notes:*

n.a. = not available;

a  not comparable with OECD data in Table 5.7 due to differences in reporting system since 1993.

GSP, Lomé Agreements, etc.). Even in the absence of trade restrictions, a German bank or tourist agency, for example, will not be able to substitute its branch offices in East Asia or Latin America with those in Hungary or Romania. Thus, German investors have hardly any chance of diverting their existing or future FDI from developing countries to CEE in the services sector. This sector now absorbs two thirds of total stock of German FDI (Table 5.7). In CEECs as well as developing countries these shares are considerably lower, but have been rising (Table 5.10). The fact that German FDI in developing countries has risen is, *ceteris paribus*, a sign against investment diversion from these countries.

In a recent survey of German enterprises,[5] Beifuß (1996) highlighted the penetration and preservation of markets as the most important motivations of their investments in CEECs (Table 5.11). This is in line with the existing literature on German and other FDI showing that market size and growth prospects of host countries are the dominant determinants of capital inflows (Agarwal *et al.*, 1991). Low wage costs in CEE were ranked at third place by German enterprises for their investment decisions in this region. They were relatively more important in the case of German investors in the Czech Republic, which attracts investments from many SMEs domiciled not far from the border. Moreover, German enterprises were apparently

*Table 5.11* Motivation of German investors in CEE (survey results)

| Rank | Reason for investment decision | Degree of importance[a] |
|------|--------------------------------|--------------------------|
| 1 | Penetration of new market | 3.22 |
| 2 | Preservation of potential market | 3.08 |
| 3 | Low wage costs | 2.76 |
| 4 | Preservation and servicing of existing markets | 2.70 |
| 5 | Competitiveness through relocation of production of intermediary products | 2.12 |
| 6 | Low tax burden | 1.46 |
| 7 | Longer working hours | 1.38 |
| 8 | Jumping of import barriers | 1.36 |

*Source*: Beifuß (1996).

*Note*:

a Investors were asked to scale their answers between 4 (most important) and 0 (unimportant). Numbers given are averages of all answering firms.

comparing their local wage costs with wage costs in the Czech Republic, which are significantly lower than the former. The comparison was not between CEECs and developing countries, which could have given an indication whether German enterprises were contemplating shifting some of their existing or potential investments from the latter to the former. German enterprises are likely to be aware that cost advantages in CEE *vis-à-vis* the German labour market are likely to be eroded in the course of convergence of CEE's economies toward EU income levels and with the resumption of realistic exchange rates. This is going to be a slow process, but unavoidable if the associated CEECs are to become full EU members. As compared to them, wage costs in most developing countries can be expected to remain comparatively low for a longer time period.

Considering German FDI in all the three sectors – natural resources, manufacturing and services of developing countries – there appears no doubt that it is overwhelmingly location bound. German FDI in developing countries in the first half of this decade has increased over-proportionately, although not as much as in CEE. The growth of German FDI in developing countries has occurred almost exclusively in Latin America (Table 5.3). Latin America, particularly Brazil, has always been a stronghold for German MNEs. The recovery of economic performance and prospects from the set-backs of the 1980s has re-encouraged the flow of German equity capital into Latin America. This process may be further strengthened by the ongoing regional integration of this continent. Regional integration tends to influence FDI inflows positively both from members and third countries (Nunnenkamp *et al.*, 1994). Africa (except South Africa) is not attractive for German investors when measured by its shares in capital outflows from Germany. It provides a good example of the belief that low wages, cheap land and lax environmental standards or preferential agreements – many African countries are members of the Lomé Convention – are not sufficient conditions to attract FDI. Large markets, strong growth prospects and good infrastructure are a good deal better determinants of it.

Many Asian developing countries offer very attractive locational conditions. But the Asian share in German outward stock of FDI has increased marginally from 2.3 per cent in 1989 to 2.6 per cent in 1994 (Deutsche Bundesbank, 1994 and 1996). In terms of outflows, it even declined in the period 1993–4 compared to 1991–2 (Table 5.3). Many German investors have been slow to respond to the growth potential of East Asian economies. During the 1960s and 1970s they mostly preferred to supply these markets through exports rather than local production in contrast to Japanese and US MNEs. The latter have established strong regional core networks of trade and local supply in East Asia. This now provides them with a competitive advantage over German and other European investors who want to come as new entrants to this market. Moreover, foreign investors are faced with increasing competition from local and intra-regional MNEs. Indeed, larger German MNEs such as Siemens or the chemical giants are very well represented in the region. It is SMEs particularly which face problems in finding new market entries.

The German government has sent several high level delegations to selected Asian countries in the past few years to bolster mutual economic cooperation through investment and trade. At the EU level, institutional arrangements (ASEM, Asia-Invest Programme) are in progress towards the same objective (EC-UNCTAD 1996: 63–86). A success of these efforts would boost German FDI in Asian countries (Agarwal, 1996b). This may be expected to lead simultaneously to a further growth of German FDI in CEE. As already discussed, several countries in CEE have strong locational advantages. This enables them to play the role of hinterland for German and other European enterprises as South and Southeast Asia do for Japan, and Latin America for the United States. So far only 2.3 per cent of German and 1.3 per cent of EU stocks of FDI are located in CEE, as compared to 16.4 per cent of Japanese and 15.5 per cent of US stock of FDI in developing Asia and Latin America, respectively (OECD, 1996). Judging from this point of view, there is great scope for further growth of German MNEs in CEE. But this need not necessarily impair the flow of German direct investments into developing countries so long as their absorptive capacity continues to expand.

## Conclusions

German MNEs have raised their share of global FDI outflows during the first half of this decade. The strongest growth in German FDI has been recorded in Central and Eastern Europe. Markets, geographical proximity, low labour costs, privatization, preferential trading arrangements and cultural relations have facilitated the involvement of German MNEs in this region.

From the perspective of developing countries, it is relevant to note that the growth of German FDI in Central and Eastern Europe has occurred not at their expense, although the possibility of investment diversion in some highly labour intensive industries cannot be completely ruled out. One apparent indication of the absence of investment diversion is that German FDI in developing countries has increased parallel to that in Central and Eastern Europe. Moreover, a diversion of German FDI is unlikely because most of it in developing countries is location-bound. Only a minor proportion of German FDI is in labour intensive branches of textiles, leather and clothing, which can be mobile between different locations. The data for Central and Eastern European countries indicate that these industries have hosted so far very small ratios of FDI inflows.

A recent survey of German MNEs recorded penetration and preservation of markets as their most important motive for investing in Central and Eastern Europe. Low wage costs and relocation of production to improve competitiveness figured at third and fifth places, respectively.

The growth of German FDI in Central and Eastern Europe does not appear to be at the expense of domestic capital formation either; but one can indeed see that many Bavarian firms have located production units in the neighbouring border areas of the Czech Republic. The local authorities of some of the regions in eastern Bavaria are worried about job losses. But for the German economy as a whole, FDI outflows have recorded declining ratios of gross fixed capital formation.

German and other European FDI is expected to rise more than proportionately in CEE. About 16 per cent of the US and Japanese stocks of FDI are invested in Latin America and Asia, respectively. Germany has so far only 2 per cent and the European Union 1 per cent of its FDI stock located in CEE. Judging on the basis of the US and Japanese geographical pattern, and of favourable prospects of growing market size in this region, German and other European MNEs can be expected to further increase their activities in CEE. The emergence of CEECs as hosts for German MNEs has indeed increased locational competition for developing countries. But this is not expected to retard the growth of German FDI in developing countries provided their absorptive capacity based on income growth, infrastructure and macroeconomic stability continues to grow.

## Notes

1 In 1993, the share of labour intensive industries, i.e., textile, leather and clothing in the EU stock of outward FDI in the world was 0.2 per cent and in developing Asia 0.1 per cent only (EC-UNCTAD, 1996: 23). In the US, the share of these industries in total outward FDI stock in 1994 amounted to 0.4 per cent. In

Japan and the UK, these shares were 1.6 per cent and 1.1 per cent respectively (OECD, 1996).

2 Specially in the clothing and textile industries, German enterprises have been subcontracting the processing of wage intensive value added activities (sewing, packing, etc.) to other countries. Some CEECs attracted these German contracts even before the beginning of their transformation process in 1988/9. Since then some of them as partners of Europe Agreements (the Czech Republic, Hungary and Portugal) have received proportionately more of such contracts. However, shifts in their shares have been at the cost of Switzerland and Austria in textiles, and of South European countries in clothing. Developing countries have attracted since then more value adding business from Germany rather than losing it to CEECs (DIW, 1995: 342).

3 In this study labour intensive industries include 'other manufactures' besides textiles, leather and clothing. Their shares increased from 1.6 per cent (1985) to 2.1 per cent (1993) in EU FDI stock, and from 1.8 per cent (1985) to 4.1 per cent (1994) in US stock of FDI in Asian developing countries.

4 They are not the same as services defined in trade statistics which include payment for shipments, cargo, insurance, trade financing, etc. For a distinction between trade related and other services see Deardorff (1984).

5 The survey included both MNEs and SMEs, who have already invested in CEE as well as those who have not yet invested there. It included investors from all the three sectors (primary, secondary and services).

# 6

# JAPANESE FOREIGN DIRECT INVESTMENT IN ASIA

## Its impact on export expansion and technology acquisition of the host economies

*Shujiro Urata*

## Introduction

The world economy has witnessed a remarkable expansion of foreign direct investment (FDI) since the early 1980s. In the post-World War II period through the 1970s, international trade grew more rapidly than FDI, and thus international trade was by far the most important international economic activity. The situation changed dramatically in the middle of the 1980s, when world FDI started to increase sharply. Indeed, for the period from the early 1980s through the mid-1990s the rate of growth of FDI was significantly higher compared to international trade. Although major investors as well as recipients have been developed countries such as the United States, Western Europe and Japan, developing countries in East Asia, including the NIEs (Hong Kong, Korea, Singapore and Taiwan), ASEAN4 (Indonesia, Malaysia, Philippines, Thailand), and China, have become large recipients of FDI.

The rapid expansion of world FDI resulted from several factors including technical progress in telecommunication services and major currency realignment. Technical progress in telecommunication services facilitates international communications involving parent companies and their overseas affiliates, while major currency realignment has provided companies with the opportunities for making profits by undertaking FDI.

The change in the attitude towards FDI in a number of countries also contributed to the rapid FDI expansion. Before the 1980s FDI was regarded as a threat to the recipient (host) country as FDI by powerful multinationals might take over the host market. This view changed in the 1980s, as the benefits of FDI such as acquisition of investable funds and technologies came to be given greater attention. Since the 1980s, attracting FDI has been one of the most important policy goals of both developing and developed countries.

To achieve the objective a number of countries have not only liberalized restrictions on FDI but also provided incentives to attract FDI.

In light of the increasing importance of Japan and developing Asia in world FDI, this chapter examines the impact of Japanese FDI on Asian countries. In particular, its impacts on exports and technology acquisition of developing Asian countries are closely analysed.

The chapter is organized as follows. The next section provides the recent trends of Japanese FDI with an emphasis on Asia, and the third section examines the motives of Japanese FDI in Asia. These two sections set the stage for the subsequent analysis of the impact of Japanese FDI on the exports and technology acquisition of Asian countries, which are respectively carried out in the fourth and fifth sections. The last section presents some concluding comments.

## Recent trends in Japanese foreign direct investment[1]

### *Rapid expansion of Japanese FDI in the 1980s*

Japanese FDI increased on a large scale and underwent major changes in its regional and sectoral composition in the latter half of the 1980s (Figures 6.1 and 6.2 and Table 6.1). The scale of FDI during the four year period of 1986–9 was unprecedented, far exceeding the total FDI from all previous years combined. Equally as dramatic as the size of the boom was the pace at which Japanese FDI declined after reaching a peak in 1989. The decline in annual FDI continued until 1992, when the magnitude of annual FDI amounted to almost one half of that recorded in 1989. In 1993 Japanese FDI started to rise again, and the upward trend is continuing.

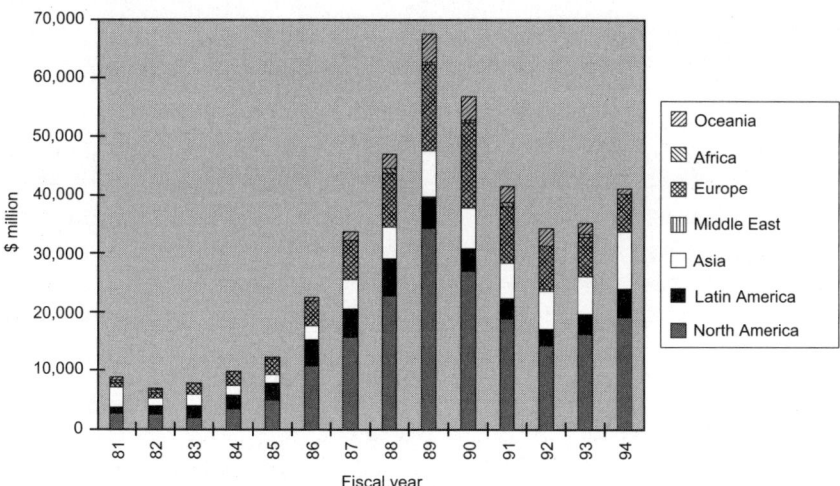

*Figure 6.1* Japanese foreign direct investment by region
*Source*: Japan, Ministry of Finance.

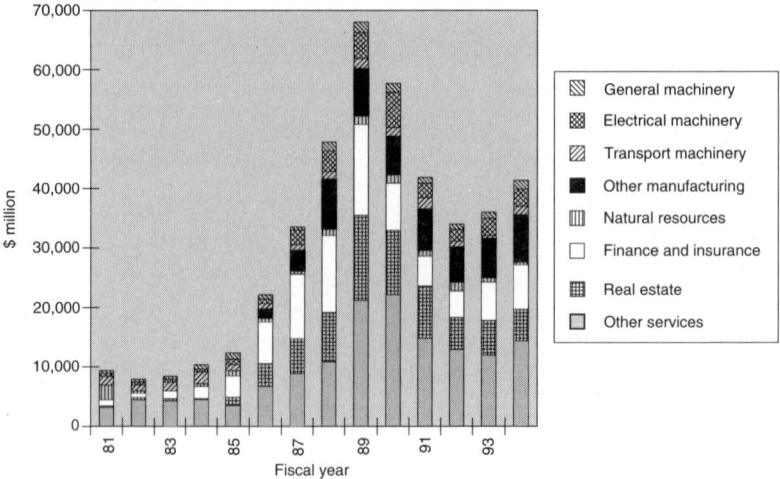

*Figure 6.2* Japanese foreign direct investment by sector
*Source*: Japan, Ministry of Finance.

One identifies both 'push' and 'pull' factors behind the recent develop-ments of Japanese FDI. The push factors are those in the investing country, that is Japan in this case, while the pull factors are those in the recipient coun-tries. We will take up the push factors in this section and the pull factors in turn in the next section in explaining the recent patterns of Japanese FDI.

Several push factors were responsible for the rapid growth of Japanese FDI in the latter half of the 1980s. The rapid and steep appreciation of the yen exchange rate against other currencies was the most important macro-economic factor leading to the expansion of Japanese FDI in the second half of the 1980s. The yen appreciated by 37 per cent between 1985 and 1988 on a real, effective basis. The drastic yen appreciation stimulated Japanese FDI in two ways. One way was the dramatic 'relative price' effect, and the other was the 'liquidity' or 'wealth' effect. The relative price effect substantially reduced the international price competitiveness of Japanese products, resulting in a decline in Japan's export volume. To cope with the new international price structure, a number of Japanese manufacturing firms moved their production base to foreign countries, especially to East Asia, where production costs were lower.

Yen appreciation had a positive impact on Japanese FDI through the 'liquidity' or 'wealth' effect as well. To the extent that yen appreciation made Japanese firms relatively more 'wealthy' in the sense of increased collat-eral and liquidity, it enabled them to finance FDI relatively more cheaply than their foreign competitors. A number of FDI projects in real estate were undertaken by Japanese firms by taking advantage of the liquidity effect.

Table 6.1a  Japanese foreign direct investment, 1980–94 ($ million)

| Region/Country | 1980 | 1981 | 1982 | 1983 | 1984 | 1985 | 1986 | 1987 | 1988 | 1989 | 1990 | 1991 | 1992 | 1993 | 1994 |
|---|---|---|---|---|---|---|---|---|---|---|---|---|---|---|---|
| World | 4,693 | 8,932 | 7,703 | 8,145 | 10,155 | 12,217 | 22,320 | 33,364 | 47,022 | 67,540 | 56,911 | 41,584 | 34,138 | 36,025 | 41,051 |
| Asia | 1,186 | 3,339 | 1,385 | 1,847 | 1,628 | 1,435 | 2,327 | 4,868 | 5,569 | 8,238 | 7,054 | 5,936 | 6,425 | 6,637 | 9,699 |
| NIEs | 378 | 722 | 738 | 1,117 | 808 | 718 | 1,531 | 2,581 | 3,264 | 4,901 | 3,355 | 2,203 | 1,920 | 2,421 | 2,865 |
| Hong Kong | 156 | 329 | 400 | 563 | 412 | 131 | 502 | 1,072 | 1,662 | 1,898 | 1,785 | 925 | 735 | 1,238 | 1,133 |
| Korea | 35 | 73 | 103 | 129 | 107 | 134 | 436 | 647 | 483 | 606 | 284 | 260 | 225 | 247 | 400 |
| Singapore | 140 | 266 | 180 | 322 | 225 | 339 | 302 | 494 | 747 | 1,902 | 840 | 613 | 670 | 644 | 1,054 |
| Taiwan | 47 | 54 | 55 | 103 | 65 | 114 | 291 | 367 | 372 | 494 | 446 | 405 | 292 | 292 | 278 |
| ASEAN4 | 786 | 2,568 | 621 | 651 | 680 | 597 | 552 | 1,031 | 1,966 | 2,782 | 3,243 | 3,082 | 3,197 | 2,398 | 3,888 |
| Indonesia | 529 | 2,434 | 410 | 374 | 374 | 408 | 250 | 545 | 586 | 631 | 1,105 | 1,193 | 1,676 | 813 | 1,759 |
| Malaysia | 146 | 31 | 83 | 140 | 142 | 79 | 158 | 163 | 387 | 673 | 725 | 880 | 704 | 800 | 742 |
| Philippines | 78 | 72 | 34 | 65 | 46 | 61 | 21 | 72 | 134 | 202 | 258 | 203 | 160 | 207 | 668 |
| Thailand | 33 | 31 | 94 | 72 | 119 | 48 | 124 | 250 | 859 | 1,276 | 1,154 | 807 | 657 | 578 | 719 |
| China | 12 | 26 | 18 | 3 | 114 | 100 | 226 | 1,226 | 296 | 438 | 349 | 579 | 1,070 | 1,691 | 2,565 |
| India | 2 | 15 | 3 | 7 | 14 | 13 | 11 | 21 | 24 | 18 | 30 | 14 | 122 | 35 | 96 |
| Latin America | 588 | 1,181 | 1,503 | 1,878 | 2,290 | 2,616 | 4,737 | 4,816 | 6,428 | 5,238 | 3,628 | 3,337 | 2,726 | 3,370 | 5,231 |
| Middle East | 158 | 96 | 124 | 175 | 273 | 45 | 44 | 62 | 259 | 66 | 27 | 90 | 709 | 217 | 290 |
| Africa | 139 | 573 | 489 | 364 | 326 | 172 | 309 | 272 | 653 | 671 | 551 | 748 | 238 | 539 | 346 |
| Oceania | 448 | 424 | 421 | 191 | 157 | 525 | 992 | 1,413 | 2,669 | 4,618 | 4,166 | 3,278 | 2,406 | 2,035 | 1,432 |
| North America | 1,596 | 2,522 | 2,905 | 2,701 | 3,544 | 5,495 | 10,441 | 15,357 | 22,328 | 33,902 | 27,192 | 18,823 | 14,572 | 15,287 | 17,823 |
| Europe | 578 | 876 | 876 | 990 | 1,937 | 1,930 | 3,469 | 6,576 | 9,116 | 14,808 | 14,294 | 9,371 | 7,061 | 7,940 | 6,230 |

Table 6.1b  Japanese foreign direct investment: share (per cent)

| Region/Count | 1980 | 1981 | 1982 | 1983 | 1984 | 1985 | 1986 | 1987 | 1988 | 1989 | 1990 | 1991 | 1992 | 1993 | 1994 |
|---|---|---|---|---|---|---|---|---|---|---|---|---|---|---|---|
| World | 100 | 100 | 100 | 100 | 100 | 100 | 100 | 100 | 100 | 100 | 100 | 100 | 100 | 100 | 100 |
| Asia | 25.3 | 37.4 | 18.0 | 22.7 | 16.0 | 11.7 | 10.4 | 14.6 | 11.8 | 12.2 | 12.4 | 14.3 | 18.8 | 18.4 | 23.6 |
| NIEs | 8.1 | 8.1 | 9.6 | 13.7 | 8.0 | 5.9 | 6.9 | 7.7 | 6.9 | 7.3 | 5.9 | 5.3 | 5.6 | 6.7 | 7.0 |
| Hong Kong | 3.3 | 3.7 | 5.2 | 6.9 | 4.1 | 1.1 | 2.3 | 3.2 | 3.5 | 2.8 | 3.1 | 2.2 | 2.2 | 3.4 | 2.8 |
| Korea | 0.7 | 0.8 | 1.3 | 1.6 | 1.1 | 1.1 | 2.0 | 1.9 | 1.0 | 0.9 | 0.5 | 0.6 | 0.7 | 0.7 | 1.0 |
| Singapore | 3.0 | 3.0 | 2.3 | 4.0 | 2.2 | 2.8 | 1.4 | 1.5 | 1.6 | 2.8 | 1.5 | 1.5 | 2.0 | 1.8 | 2.6 |
| Taiwan | 1.0 | 0.6 | 0.7 | 1.3 | 0.6 | 0.9 | 1.3 | 1.1 | 0.8 | 0.7 | 0.8 | 1.0 | 0.9 | 0.8 | 0.7 |
| ASEAN4 | 16.7 | 28.8 | 8.1 | 8.0 | 6.7 | 4.9 | 2.5 | 3.1 | 4.2 | 4.1 | 5.7 | 7.4 | 9.4 | 6.7 | 9.5 |
| Indonesia | 11.3 | 27.3 | 5.3 | 4.6 | 3.7 | 3.3 | 1.1 | 1.6 | 1.2 | 0.9 | 1.9 | 2.9 | 4.9 | 2.3 | 4.3 |
| Malaysia | 3.1 | 0.3 | 1.1 | 1.7 | 1.4 | 0.6 | 0.7 | 0.5 | 0.8 | 1.0 | 1.3 | 2.1 | 2.1 | 2.2 | 1.8 |
| Philippines | 1.7 | 0.8 | 0.4 | 0.8 | 0.4 | 0.5 | 0.1 | 0.2 | 0.3 | 0.3 | 0.5 | 0.5 | 0.5 | 0.6 | 1.6 |
| Thailand | 0.7 | 0.3 | 1.2 | 0.9 | 1.2 | 0.4 | 0.6 | 0.7 | 1.8 | 1.9 | 2.0 | 1.9 | 1.9 | 1.6 | 1.8 |
| China | 0.3 | 0.3 | 0.2 | 0.0 | 1.1 | 0.8 | 1.0 | 3.7 | 0.6 | 0.6 | 0.6 | 1.4 | 3.1 | 4.7 | 6.2 |
| India | 0.0 | 0.2 | 0.0 | 0.1 | 0.1 | 0.1 | 0.0 | 0.1 | 0.1 | 0.0 | 0.1 | 0.0 | 0.4 | 0.1 | 0.2 |
| Latin America | 12.5 | 13.2 | 19.5 | 23.1 | 22.6 | 21.4 | 21.2 | 14.4 | 13.7 | 7.8 | 6.4 | 8.0 | 8.0 | 9.4 | 12.7 |
| Middle East | 3.4 | 1.1 | 1.6 | 2.1 | 2.7 | 0.4 | 0.2 | 0.2 | 0.6 | 0.1 | 0.0 | 0.2 | 2.1 | 0.6 | 0.7 |
| Africa | 3.0 | 6.4 | 6.3 | 4.5 | 3.2 | 1.4 | 1.4 | 0.8 | 1.4 | 1.0 | 1.0 | 1.8 | 0.7 | 1.5 | 0.8 |
| Oceania | 9.5 | 4.7 | 5.5 | 2.3 | 1.5 | 4.3 | 4.4 | 4.2 | 5.7 | 6.8 | 7.3 | 7.9 | 7.0 | 5.6 | 3.5 |
| North America | 34.0 | 28.2 | 37.7 | 33.2 | 34.9 | 45.0 | 46.8 | 46.0 | 47.5 | 50.2 | 47.8 | 45.3 | 42.7 | 42.4 | 43.4 |
| Europe | 12.3 | 9.8 | 11.4 | 12.2 | 19.1 | 15.8 | 15.5 | 19.7 | 19.4 | 21.9 | 25.1 | 22.5 | 20.7 | 22.0 | 15.2 |

Source: Japan, Ministry of Finance (unpublished data).
Notes: Reported values for the fiscal year, which starts in April and ends in March in the following year.

Another important push factor was the emergence of the 'bubble' economy in Japan. Indeed, the liquidity effect discussed above was strengthened by the bubble economy, in which the prices of assets such as shares and land increased enormously. Indeed, the average share prices more than doubled in four years from 1985 to 1989, as the index of share prices increased from 45.7 in 1985 to 117.8 in 1989. The Bank of Japan injected liquidity into the economy to deal with the recessionary impact caused by the drastic yen appreciation. Active fiscal spending for the same purpose of reflating the economy was an additional factor leading to the bubble economy.

A general rise in Japanese firms' managerial and technological capabilities in international business, which had been accumulated through past experiences such as exporting and FDI, was a natural underlying factor behind the surge of Japanese FDI. It is also important to note that a number of Japanese firms followed their business customers that invested overseas. A case in point is FDI by subcontracting firms that followed their parents, which had undertaken FDI, to maintain business. Furthermore, the labour shortage in Japan necessitated some Japanese firms, especially small and medium sized firms, to move their operation abroad.

The decline in Japanese FDI in the early 1990s resulted mainly from the bursting of the bubble economy in 1989. The depreciation of the yen also contributed to the decline. The recent increase in FDI is attributable to the dramatic appreciation of the yen, whose real effective exchange rate increased by 17 per cent in a year from 1992 to 1993. The appreciation of the yen set off the relative price effect discussed above, to result in the expansion of Japanese FDI. It is worth noting that unlike the dramatic increase in Japanese FDI in the second half of the 1980s, the recent increase in Japanese FDI is taking place in the absence of the bubble economy. This difference is reflected in the changes in the regional as well as sectoral distribution of Japanese FDI over time, to which we will turn next.

## Japanese foreign direct investment in Asia

Japanese FDI in the second half of the 1980s was directed largely to North America and Europe, mainly in non-manufacturing sectors such as services and real estate. These two developed regions together absorbed two-thirds of Japanese FDI outflows during the period. Although a smaller share of Japanese FDI went to Asia, in the 1980s investments in manufacturing were relatively active. The 1990s have seen some changes in the patterns of Japanese FDI. First, the share of Asia, particularly East Asia including the NIEs, ASEAN countries, China and Vietnam, in Japanese FDI started to increase (Table 6.1b). Major pull factors behind Japanese FDI in East Asia include the region's robust economic growth, low unit labour costs, and trade and FDI liberalization and pro-FDI policy. Although Japanese FDI to South Asian countries such as India started to increase in the 1990s, its share is still quite small compared to the share for East Asia. Another noticeable change was that Japanese manufacturing firms have been undertaking FDI actively, particularly in East Asia.

Since the mid-1980s, the geographical distribution of Japanese FDI to Asia has changed significantly, from the Asian NIEs (Newly Industrializing Economies) to ASEAN (Association of South East Asian Nations), and then to China and other Asian countries.[2] The NIEs attracted FDI until the late 1980s through FDI promotion policy. Policy makers in Korea, Taiwan, and Singapore, in particular, promoted inward FDI in their pursuit of high-tech industrialization. These countries enjoyed positive growth brought about by the simultaneous expansion of trade and inward FDI.

However, Japanese FDI in the Asian NIEs reached a peak in the late 1980s just as its overall FDI also peaked. The Asian NIEs started to lose some of their cost advantages due to rapid wage increases and currency appreciation. Firms in Japan and other advanced economies therefore started to look to other East Asian countries such as ASEAN as hosts for investment. One important factor in attracting FDI in manufacturing to ASEAN has been the ASEAN countries' shift from inward-oriented to outward-oriented strategies, which were carried out through their unilateral liberalization of trade and FDI policy. Such regime changes have been prompted by the earlier success of outward-oriented policy in the NIEs.

FDI inflows into China have also grown quickly since 1988–9 due to China's gradual but persistent economic reforms, liberalization in trade and FDI policy, and its political and social stability despite the Tiananmen Square incident in 1989. As of 1994, China was the largest recipient of Japanese FDI in Asia. The attractiveness of China as a host to FDI has increased recently because some ASEAN countries have lost their attractiveness mainly due to the rapid increase in the production costs including wages, material and services costs, which in turn resulted from currency appreciation, shortage of manpower, the emergence of serious bottlenecks in infrastructure and other factors.[3]

In recent years, although on a significantly lower scale, Japanese FDI to other Asian countries such as Vietnam and India has begun to increase. Liberalization of FDI policy by these countries attracted Japanese investors. Japanese FDI to these countries is likely to increase, as a recent questionnaire survey of 264 Japanese companies conducted by the Export-Import Bank of Japan revealed that India and Vietnam ranked second and third, respectively, behind China as attractive host countries to Japanese investors.[4] However, the same survey points out that these countries have to improve their infrastructure in order to successfully attract Japanese FDI.

Since the 1980s, Japanese FDI in East Asia has primarily been aimed at the manufacturing and services sectors. A distinct characteristic of Japanese FDI in Asia is the relatively large share of manufacturing in comparison to other regions. The magnitude of Japanese manufacturing FDI increased more than ten-fold in nine years from $0.5 billion in 1985 to $5.2 billion in 1994. The rate of increase in Japanese manufacturing FDI in East Asia during the period was greater than the corresponding FDI to the world, and the share of Asia in Japanese manufacturing FDI rose from 19.6 per cent in 1985 to 32.9 per cent in 1994. The share of manufacturing in overall Japanese FDI in Asia also rose from 32.1 per cent to 55.1 per cent over the 1985–94 period. The geograph-ical shift in the importance of Japanese manufacturing FDI away from the NIEs to ASEAN and then to China is pronounced.

Sectoral composition of Japanese manufacturing FDI in Asia has changed notably since the latter half of the 1980s. The most remarkable development is that Japanese FDI in electrical machinery (including electronics) has expanded sharply (Table 6.2). Until the mid-1980s, Japanese FDI in electrical machinery in Asia had been relatively small compared to other sectors. Japanese FDI in chemical products, general machinery and transport machinery in the Asian NIEs was relatively greater than Japanese FDI in electrical machinery in the same area, and Japanese FDI in transport machinery, iron and steel and textiles in the ASEAN economies was larger than in electrical machinery. However, since the mid-1980s electrical machinery has become the sector with the most active FDI. Japanese FDI in China's electrical machinery industry was also the largest between the mid-1980s and the early 1990s. It is interesting to observe that the largest recipient of Japanese FDI in elec-trical machinery changed from the NIEs in 1986–8 to ASEAN in 1989–92 and to China in 1993 (figures are not shown in the table). Although there are signs that other Asian countries such as India and Vietnam started to attract Japanese FDI in electrical machinery, its magnitude is still quite low.

## Motives behind Japanese FDI in Asia

Several factors promoting recent Japanese FDI were explained in the previous section. This section examines the motives behind Japanese FDI in more detail, because these motives influence the impact of Japanese FDI on the recipient countries, which is one of our main concerns in this chapter. Besides, an examination of the motives of Japanese FDI would pro-vide useful information to the policy makers engaged in the formulation of FDI policy.

Several questionnaire surveys on the motives of Japanese firms for under-taking FDI are conducted on a regular basis. The most comprehensive is a questionnaire survey conducted by the Ministry of International Trade and Industry (MITI) of the Japanese government.[5] Table 6.3 summarizes the results of the 1992 MITI survey conducted on the motives for Japanese FDI in Asia, Latin America and the world as a whole. One common motive behind Japanese FDI in the three regions is the importance of local sales. Indeed, as many as 52–66 per cent of the respondents indicated local sales as one of

Table 6.2 Japanese foreign direct investment in manufacturing in Asia, 1983–94 ($million)

(a) 1983–54 (cumulative)

| Region/Country | Food | Textiles | Wood and pulp | Chemical products | Ferrous and non-ferrous | General machinery | Electrical machinery | Transport machinery | Others | Total manufacturing |
|---|---|---|---|---|---|---|---|---|---|---|
| Asia | 80 | 179 | 29 | 302 | 210 | 213 | 189 | 340 | 173 | 1,714 |
| East Asia | 80 | 179 | 28 | 302 | 209 | 213 | 188 | 306 | 154 | 1,658 |
| NIEs | 28 | 11 | 3 | 241 | 29 | 169 | 117 | 127 | 90 | 815 |
| Hong Kong | 10 | 3 | 0 | 4 | – | 2 | 8 | – | 5 | 32 |
| Korea | 6 | 5 | 1 | 10 | 4 | 13 | 27 | 27 | 13 | 105 |
| Singapore | 10 | 1 | 1 | 208 | 11 | 135 | 45 | 4 | 21 | 435 |
| Taiwan | 2 | 2 | 2 | 20 | 13 | 20 | 37 | 96 | 51 | 243 |
| ASEAN4 | 37 | 166 | 24 | 56 | 176 | 40 | 66 | 179 | 56 | 799 |
| Indonesia | 4 | 118 | 15 | 29 | 75 | 3 | 11 | 57 | 22 | 335 |
| Malaysia | 3 | 17 | 6 | 6 | 88 | 3 | 31 | 83 | 22 | 258 |
| Philippines | 11 | – | 0 | 6 | 7 | 0 | 2 | 35 | 2 | 64 |
| Thailand | 19 | 31 | 2 | 15 | 5 | 33 | 22 | 5 | 10 | 142 |
| China | 14 | 2 | 2 | 6 | 4 | 3 | 4 | 1 | 8 | 44 |
| India | 0 | 0 | 0 | 0 | 0 | 0 | 2 | 31 | 0 | 33 |

Table 6.2  (Contd.)

(b) 1986–94 (cumulative)

| Region/Country | Food | Textiles | Wood and pulp | Chemical products | Ferrous and non-ferrous | General machinery | Electrical machinery | Transport machinery | Others | Total manufacturing |
|---|---|---|---|---|---|---|---|---|---|---|
| Asia | 1,523 | 1,925 | 564 | 4,328 | 2,448 | 2,368 | 7,013 | 2,033 | 3,809 | 26,013 |
| East Asia | 1,512 | 1,815 | 561 | 4,161 | 2,392 | 2,358 | 6,904 | 1,904 | 3,587 | 25,198 |
| NIEs | 749 | 266 | 68 | 1,114 | 641 | 676 | 1,963 | 543 | 1,018 | 7,041 |
| Hong Kong | 95 | 58 | 4 | 41 | 66 | 169 | 419 | 48 | 257 | 1,157 |
| Korea | 85 | 45 | 9 | 223 | 110 | 167 | 393 | 289 | 153 | 1,474 |
| Singapore | 473 | 131 | 32 | 584 | 98 | 175 | 577 | 12 | 263 | 2,345 |
| Taiwan | 97 | 32 | 24 | 266 | 367 | 166 | 574 | 194 | 345 | 2,066 |
| ASEAN4 | 472 | 742 | 427 | 2,779 | 1,433 | 1,134 | 3,553 | 1,005 | 1,758 | 13,306 |
| Indonesia | 111 | 402 | 274 | 1,915 | 342 | 83 | 286 | 302 | 180 | 3,898 |
| Malaysia | 25 | 28 | 92 | 420 | 394 | 362 | 1,517 | 150 | 949 | 3,937 |
| Philippines | 46 | 27 | 10 | 102 | 148 | 28 | 469 | 222 | 126 | 1,180 |
| Thailand | 289 | 285 | 51 | 342 | 549 | 661 | 1,281 | 330 | 503 | 4,291 |
| China | 291 | 807 | 65 | 268 | 318 | 548 | 1,388 | 356 | 812 | 4,852 |
| India | 4 | 5 | 0 | 69 | 7 | 4 | 92 | 70 | 68 | 318 |

Source: Japan, Ministry of Finance.
Notes: Reported values for the fiscal year, which starts in April and ends in March in the following year.

Table 6.3  Motives behind Japanese foreign direct investment in Asia and Latin America, 1992

| Sector | World | | | | | Asia | | | | | Latin America | | | | |
|---|---|---|---|---|---|---|---|---|---|---|---|---|---|---|---|
| | Local sales | Exports to third countries | Exports to Japan | Use of local labor | FDI promotion policy by hosts | Local sales | Exports to third countries | Exports to Japan | Use of local labor | FDI promotion policy by hosts | Local sales | Exports to third countries | Exports to Japan | Use of local labor | FDI promotion policy by hosts |
| Total industry | 66.1 | 21.1 | 10.7 | 23.7 | 10.8 | 65.6 | 26.9 | 16.2 | 45.4 | 18.3 | 52.5 | 17.2 | 8.9 | 27.3 | 18.8 |
| Manufacturing | 69.6 | 22.3 | 14.6 | 43.6 | 21.5 | 62.4 | 25.5 | 21.1 | 65.0 | 28.0 | 68.1 | 15.3 | 6.9 | 52.8 | 40.3 |
| Food | 50.0 | 22.5 | 26.1 | 29.6 | 5.6 | 53.2 | 30.6 | 27.4 | 48.4 | 9.7 | 36.4 | 45.5 | 18.2 | 54.5 | 9.1 |
| Textiles | 55.6 | 31.6 | 31.0 | 69.6 | 18.7 | 50.8 | 32.3 | 37.9 | 80.6 | 17.7 | 63.6 | 22.7 | 13.6 | 63.6 | 22.7 |
| Wood and pulp | 33.3 | 2.8 | 33.3 | 33.3 | 11.1 | 20.0 | 6.7 | 33.3 | 46.7 | 6.7 | 0.0 | 0.0 | 0.0 | 50.0 | 0.0 |
| Chemical products | 75.8 | 18.9 | 10.1 | 29.0 | 15.5 | 77.7 | 20.2 | 12.4 | 44.6 | 21.2 | 55.6 | 0.0 | 11.1 | 11.1 | 22.2 |
| Iron and steel | 84.6 | 9.9 | 4.4 | 41.8 | 6.6 | 82.7 | 13.5 | 5.8 | 59.6 | 5.8 | 75.0 | 0.0 | 0.0 | 25.0 | 25.0 |
| Non-ferrous metals | 77.2 | 16.3 | 15.2 | 46.7 | 28.3 | 82.6 | 23.9 | 17.4 | 76.1 | 43.5 | 60.0 | 0.0 | 20.0 | 80.0 | 40.0 |
| General machinery | 73.6 | 20.1 | 4.2 | 39.3 | 20.9 | 64.4 | 26.4 | 9.2 | 64.4 | 24.1 | 85.7 | 28.6 | 0.0 | 42.9 | 71.4 |
| Electrical machinery | 71.0 | 25.2 | 13.8 | 43.6 | 23.5 | 59.6 | 28.2 | 20.6 | 70.6 | 32.0 | 77.5 | 5.0 | 5.0 | 47.5 | 55.0 |
| Transport machinery | 74.1 | 15.2 | 12.6 | 41.7 | 29.9 | 68.8 | 15.0 | 18.8 | 65.6 | 40.0 | 66.7 | 11.1 | 5.6 | 66.7 | 33.3 |
| Precision instruments | 66.7 | 29.3 | 14.7 | 32.0 | 10.7 | 45.5 | 27.3 | 27.3 | 60.6 | 24.2 | 66.7 | 0.0 | 0.0 | 66.7 | 0.0 |
| Petro and coal products | 50.0 | 12.5 | 25.0 | 25.0 | 0.0 | 50.0 | 0.0 | 25.0 | 50.0 | 0.0 | 100 | 0.0 | 0.0 | 0.0 | 0.0 |
| Other manufacturing | 67.5 | 27.1 | 17.1 | 55.6 | 27.3 | 55.2 | 31.2 | 26.8 | 69.2 | 34.0 | 75.0 | 25.0 | 0.0 | 75.0 | 56.3 |

Source: Japan, MITI, Kaigai Toshi Tokei Soran [Comprehensive Statistics on Japanese Foreign Direct Investment], no. 5, 1994.
Note: The figures indicate the percentage share of the firms indicating the motive in total number of surveyed firms.

their motives for undertaking FDI in these regions. For the Asian affiliates, local sales was an important motive particularly for the sectors producing the materials to be used for further processing such as chemical products, iron and steel and non-ferrous metals.

Compared to the case for the affiliates in other regions, for the affiliates in Asia the export motive captured a relatively large share; approximately one out of four affiliates responded that exporting to the third countries was one of their motives, while one out of five affiliates indicated that exporting to Japan was one of their motives. One would therefore expect that Japanese FDI in Asia would tend to stimulate exports from host countries.

A large number of respondents identified the use of local labor as their motive for FDI in Asia. This observation suggests that Japanese firms have shifted their production base to Asia in order to produce labour-intensive products for exports. Japan has long lost its comparative advantage in labour-intensive products, so Japanese firms are taking advantage of abundant labour in the Asian host countries. Japanese FDI in textiles may be a typical example reflecting such an FDI strategy on the part of Japanese firms. Indeed, as many as 80 per cent of respondents in textiles noted the use of local labour as their motive for FDI in Asia.

Despite the fact that machinery products are capital- and technology-intensive, a large proportion of the respondents in the machinery sectors undertook FDI in Asia to take advantage of local labor. This seemingly inconsistent behaviour of Japanese firms may be explained by their strategy of breaking up the entire production process into several sub-processes and locating labour-intensive sub-processes in labour abundant Asia countries. Such an arrangement gives rise to a production system under which a vertical division of labour is pursued internationally within a firm, leading to the emergence of inter-process, intra-firm, and intra-industry trade.

The use of local labour as a motive for Japanese FDI in Asia may reflect rational behaviour on the part of Japanese firms from the point of view of comparative advantage. Richly endowed with labor, Asian countries have a comparative advantage in production that requires labour-intensive techniques. This observation points to the differences between the expectation of the host countries on the one hand and the actual behaviour of foreign firms on the other regarding the technologies utilized in the host countries. The host countries interested in upgrading their technological capability would like foreign firms to bring in high technologies with a hope that these countries would gain a comparative advantage in high value added production in the future. However, foreign companies are generally interested in exploiting the comparative advantages currently possessed by the host countries. Further discussion on technology related issues will be presented in the fourth section, pp. 158–65.

FDI promotion policies by the host government appear to have played an important role in attracting Japanese FDI in Asia, since one out of four

Japanese firms in manufacturing undertook FDI in response to the incentives given to FDI. It should be noted that FDI promotion policies by host countries are important for attracting FDI in developing countries in general, not only in Asia but also in Latin America (Table 6.3). The high response rate observed for the transport machinery sector in Asia reflects the fact that automobile production, a major part of that sector, has been a target of industrial promotion policy in a number of countries in Asia including Malaysia, Thailand and Indonesia. The promotion of the transport industry has been carried out through various means including incentives given to foreign firms. In discussing the impact of FDI policy on FDI, one should consider not only FDI promotion policy but also the restrictions imposed on FDI. This is because it is the net effect of these policies that would influence FDI.

## Japanese FDI and technology exports in Asia

Foreign direct investment has been an important means for obtaining technologies for developing countries. Non-equity type arrangements such as technology trade, OEM (original equipment manufacturer) and other forms of international subcontracting have also been an effective means of obtaining technologies for developing countries.[6] Under the OEM and other subcontracting arrangements technologies are transferred from OEM order placing firms (subcontractors) to OEM receiving firms (subcontractees), as product designs and drawings are transferred as a part of these agreements.[7] Indeed, Japan and Korea relied mainly on technology imports and OEM but not on FDI for acquiring technologies from developed countries.[8]

### Recent trends in Japan's technology exports

Japan's technology exports have been on the rise in recent years with sizeable year-to-year fluctuations. The number of technology export cases increased from 1,824 in 1985 to 2,148 in 1995, and the receipt of royalties increased from $381 million to $600 million during the same period (Table 6.4). Japan's technology exports to Asia, particularly to ASEAN4, increased notably, compared to those destined for other parts of the world. In terms of royalty receipts, Japan's technology exports to Asia almost tripled from 1985 to 1995, significantly higher than the 60 per cent growth registered for Japan's technology exports overall. Japan's technology exports to ASEAN4 grew more than eight-fold in terms of royalty receipts over the same period.

Among the East Asian countries Korea has been by far the largest recipient of Japan's technology exports. In terms of the number of cases Korea received 3,875 licences from 1985 to 1995, followed by Taiwan and China at 3,875 and 2,314 licences, respectively. Among the ASEAN4 countries, Indonesia, Malaysia and Thailand respectively imported around 900 licences during

*Table 6.4*   Japan's technology exports, 1985–95

*(a) Number of cases*

| Region/Country | 1985 | 1986 | 1987 | 1988 | 1989 | 1990 | 1991 | 1992 | 1993 | 1994 | 1995 | 1985–95 |
|---|---|---|---|---|---|---|---|---|---|---|---|---|
| World | 1,824 | 2,099 | 1,730 | 1,655 | 1,850 | 2,086 | 1,570 | 2,066 | 1,983 | 1,896 | 2,148 | 20,907 |
| Asia | 970 | 1,248 | 956 | 944 | 1,054 | 1,166 | 945 | 1,402 | 1,248 | 1,160 | 1,537 | 12,630 |
| NIEs | 397 | 531 | 463 | 450 | 563 | 539 | 428 | 831 | 618 | 392 | 692 | 5,904 |
| Korea | 205 | 360 | 296 | 290 | 284 | 403 | 274 | 674 | 419 | 247 | 423 | 3,875 |
| Taiwan | 140 | 163 | 159 | 149 | 236 | 113 | 132 | 112 | 143 | 127 | 239 | 1,713 |
| Singapore | 52 | 8 | 8 | 11 | 43 | 23 | 22 | 45 | 56 | 18 | 30 | 316 |
| ASEAN4 | 199 | 306 | 132 | 196 | 279 | 244 | 325 | 276 | 321 | 392 | 428 | 3,098 |
| Indonesia | 50 | 49 | 39 | 89 | 111 | 70 | 187 | 81 | 45 | 139 | 115 | 975 |
| Malaysia | 48 | 224 | 20 | 26 | 26 | 49 | 46 | 105 | 113 | 83 | 157 | 897 |
| Phlippines | 48 | 3 | 39 | 10 | 17 | 16 | 17 | 23 | 25 | 46 | 39 | 283 |
| Thailand | 53 | 30 | 34 | 71 | 125 | 109 | 75 | 67 | 138 | 124 | 117 | 943 |
| China | 285 | 232 | 237 | 171 | 96 | 241 | 76 | 188 | 214 | 280 | 294 | 2,314 |
| India | 46 | 72 | 75 | 52 | 67 | 63 | 36 | 73 | 36 | 50 | 53 | 623 |
| Others | 43 | 107 | 49 | 75 | 49 | 79 | 80 | 34 | 59 | 46 | 70 | 691 |

*(b) Value ($ millions)*

| Region/Country | 1985 | 1986 | 1987 | 1988 | 1989 | 1990 | 1991 | 1992 | 1993 | 1994 | 1995 | 1985–95 |
|---|---|---|---|---|---|---|---|---|---|---|---|---|
| World | 381 | 435 | 358 | 350 | 343 | 460 | 435 | 555 | 519 | 614 | 600 | 5,051 |
| Asia | 135 | 185 | 154 | 137 | 174 | 185 | 258 | 363 | 275 | 382 | 375 | 2,622 |
| NIEs | 40 | 41 | 48 | 66 | 80 | 82 | 107 | 182 | 113 | 192 | 186 | 1,138 |
| Korea | 23 | 32 | 35 | 45 | 50 | 69 | 91 | 117 | 54 | 139 | 103 | 758 |
| Taiwan | 6 | 8 | 11 | 18 | 15 | 6 | 13 | 33 | 33 | 52 | 44 | 238 |
| Singapore | 11 | 1 | 3 | 4 | 15 | 7 | 2 | 32 | 26 | 2 | 39 | 142 |
| ASEAN4 | 11 | x | 15 | 32 | 40 | 60 | 111 | 136 | 78 | 124 | 91 | 698 |
| Indonesia | 4 | 19 | 3 | 10 | 12 | 15 | 73 | 78 | 13 | 27 | 30 | 283 |
| Malaysia | 4 | 1 | 6 | 3 | 5 | 16 | 10 | 48 | 39 | 36 | 23 | 192 |
| Phlippines | 0 | x | 1 | 3 | 4 | 4 | 5 | 2 | 6 | 10 | 5 | 38 |
| Thailand | 3 | 5 | 5 | 16 | 19 | 24 | 24 | 8 | 22 | 51 | 33 | 210 |
| China | 71 | 92 | 74 | 24 | 30 | 22 | 19 | 30 | 70 | 52 | 68 | 551 |
| India | 10 | 20 | 9 | 6 | 14 | 17 | 11 | 13 | 9 | 9 | 17 | 136 |
| Others | 3 | – | 8 | 8 | 10 | 5 | 10 | 1 | 3 | 5 | 14 | 98 |

*Source*: Japan, Management and Coordination Agency, Kagakugijutu Kenkyu Chosa (Survey on Science and Technology Research), various issues.

*Note*: Technology exports in this table refer to only those newly licensed. As such those continuing are not considered.

the 1985–95 period, while the Philippines imported less than 300 licences over the same period.

Turning to the sectoral composition of Japan's technology exports to Asian countries, one finds that electrical machinery has been most active in exporting technologies through technology trade (Table 6.5). Electric machinery is followed by chemical products, general machinery and transport machinery. Among the Asian countries, Korea and Taiwan have been major recipients of Japanese technologies in a number of industries. However, there are some special patterns observed for specific countries and sectors. One notable pattern is found for transport machinery, in which Thailand, Indonesia and Malaysia have been active in importing Japanese technologies. This pattern clearly reflects the automobile sector promotion policy enthusiastically pursued by these governments.

Despite the increase of Japan's technology exports to the world and especially the substantial increase of those to Asia, their rates of increase have been lower compared to Japan's FDI. Japanese FDI to the world and to Asia in terms of value respectively increased threefold and fourfold between 1985 and 1994 (Table 6.1), while the rates of increase for Japan's technology exports to the world and to Asia were significantly lower at 1.6 times and 2.8 times respectively during the same period (Table 6.4).

The increasing importance of FDI over technology exports by Japanese firms is due to the factors related to Japanese firms, the providers of technologies, and those related to Asian countries, the recipients of technologies. Concerning the factors related to Japanese firms, the reduction in the cost of undertaking FDI was most important. The drastic yen appreciation and the emergence of the bubble reduced the cost of undertaking FDI by Japanese firms. The increasing importance of the brand names of their products has made the Japanese firms put a greater emphasis on FDI rather than on technology exports as a means to extract the rents accrued from the technologies they own. Typical examples are electronic products, which have come to be identified by company names such as Sony, Panasonic and Toshiba.

As for the factors in Asian countries, a shift in the policy towards FDI from protection to promotion was important in attracting FDI, as discussed above. Furthermore, FDI has been a preferred mode of receiving technologies as FDI brings in not only the technologies but also scarce foreign exchange. By contrast, technology imports incur the foreign exchange loss to the Asian countries.

### Intra-firm technology transfer

The increasing importance of FDI in comparison with technology exports as a means of transmitting technologies from Japanese firms to Asia can also be seen from the increasing importance of intra-firm transactions in technology exports of Japanese firms. According to the results of a MITI survey shown in Table 6.6, for technology licensing to Asia by Japanese firms the proportion of intra-firm technology licensing increased from 35.9 per cent in 1986 to 46.3 per cent in 1992. Among the subsectors, intra-firm technology licensing is particularly important in electrical machinery, since as much as 71 per cent of technology licensing in that sector took the form of intra-firm licensing in 1992, supporting a hypothesis that firm specific technology including brand names in electrical machinery is best utilized inside the firms.

Table 6.5  Japan's technology exports to Asia by industry and by country, 1985–95

(a) Number of cases

| Sector | World | Asia | NIEs | Korea | Taiwan | Singapore | ASEAN4 | Indonesia | Malaysia | Philippines | Thailand | China | India | Others |
|---|---|---|---|---|---|---|---|---|---|---|---|---|---|---|
| All industries | 20,907 | 12,630 | 5,904 | 3,875 | 1,713 | 316 | 3,098 | 975 | 897 | 283 | 943 | 2,314 | 623 | 691 |
| Manufacturing | 18,619 | 11,113 | 5,180 | 3,342 | 1,604 | 234 | 2,607 | 738 | 815 | 215 | 839 | 2,194 | 588 | 544 |
| Food products | 567 | 421 | 149 | 119 | 22 | 8 | 77 | 41 | 8 | 4 | 24 | 159 | 29 | 7 |
| Textiles | 880 | 667 | 146 | 110 | 36 | 0 | 291 | 155 | 7 | 1 | 128 | 183 | 10 | 37 |
| Paper and pulp | 121 | 71 | 32 | 12 | 17 | 3 | 28 | 4 | 6 | 0 | 18 | 7 | 2 | 2 |
| Publishing | 59 | 23 | 17 | 13 | 3 | 1 | 3 | 0 | 0 | 1 | 2 | 2 | 0 | 1 |
| Chemical prdoucts | 2,782 | 1,498 | 782 | 542 | 193 | 47 | 351 | 96 | 69 | 17 | 169 | 249 | 76 | 40 |
| Petroleum products | 85 | 34 | 25 | 14 | 8 | 3 | 3 | 1 | 1 | 0 | 1 | 5 | 0 | 1 |
| Plastics | 353 | 228 | 127 | 91 | 26 | 10 | 57 | 6 | 33 | 1 | 17 | 33 | 6 | 5 |
| Rubber products | 242 | 147 | 77 | 38 | 25 | 14 | 22 | 6 | 6 | 1 | 8 | 40 | 8 | 0 |
| Ceramics | 812 | 536 | 271 | 175 | 84 | 12 | 125 | 25 | 20 | 2 | 60 | 75 | 46 | 19 |
| Iron and steel | 2,271 | 586 | 235 | 77 | 143 | 15 | 154 | 64 | 35 | 20 | 32 | 106 | 87 | 4 |
| Non-ferrous metals | 618 | 390 | 195 | 100 | 75 | 20 | 114 | 15 | 31 | 23 | 40 | 34 | 31 | 16 |
| Metal products | 793 | 640 | 377 | 305 | 68 | 4 | 181 | 41 | 71 | 28 | 19 | 70 | 4 | 8 |
| General machinery | 2,549 | 1,896 | 1,202 | 748 | 445 | 9 | 221 | 53 | 63 | 50 | 91 | 381 | 49 | 43 |
| Electrical machinery | 3,827 | 2,388 | 1,012 | 629 | 305 | 78 | 386 | 85 | 161 | 14 | 119 | 550 | 126 | 314 |
| Transport machinery | 1,796 | 1,044 | 337 | 250 | 86 | 1 | 483 | 74 | 277 | 21 | 105 | 116 | 99 | 9 |
| Precision instruments | 551 | 352 | 124 | 64 | 55 | 5 | 74 | 60 | 13 | 27 | 1 | 133 | 12 | 9 |
| Other manufacturing | 313 | 192 | 72 | 55 | 13 | 4 | 37 | 12 | 14 | 6 | 5 | 51 | 3 | 29 |

Table 6.5 (Contd.)

(b) Sectoral share (%)

| | | | | | | | | | | | | | | |
|---|---|---|---|---|---|---|---|---|---|---|---|---|---|---|
| All industries | 100 | 100 | 100 | 100 | 100 | 100 | 100 | 100 | 100 | 100 | 100 | 100 | 100 | 100 |
| Manufacturing | 89.1 | 88.0 | 87.7 | 86.2 | 93.6 | 74.1 | 84.2 | 75.7 | 90.9 | 76.0 | 89.0 | 94.8 | 94.4 | 78.7 |
| Food products | 2.7 | 3.3 | 2.5 | 3.1 | 1.3 | 2.5 | 2.5 | 4.2 | 0.9 | 1.4 | 2.5 | 6.9 | 4.7 | 1.0 |
| Textiles | 4.2 | 5.3 | 2.5 | 2.8 | 2.1 | 0.0 | 9.4 | 15.9 | 0.8 | 0.4 | 13.6 | 7.9 | 1.6 | 5.4 |
| Paper and pulp | 0.6 | 0.6 | 0.5 | 0.3 | 1.0 | 0.9 | 0.9 | 0.4 | 0.7 | 0.0 | 1.9 | 0.3 | 0.3 | 0.3 |
| Publishing | 0.3 | 0.2 | 0.3 | 0.3 | 0.2 | 0.3 | 0.1 | 0.0 | 0.0 | 0.4 | 0.2 | 0.1 | 0.0 | 0.1 |
| Chemical prdoucts | 13.3 | 11.9 | 13.2 | 14.0 | 11.3 | 14.9 | 11.3 | 9.8 | 7.7 | 6.0 | 17.9 | 10.8 | 12.2 | 5.8 |
| Petroleum products | 0.4 | 0.3 | 0.4 | 0.4 | 0.5 | 0.9 | 0.1 | 0.1 | 0.1 | 0.0 | 0.1 | 0.2 | 0.0 | 0.1 |
| Plastics | 1.7 | 1.8 | 2.2 | 2.3 | 1.5 | 3.2 | 1.8 | 0.6 | 3.7 | 0.4 | 1.8 | 1.4 | 1.0 | 0.7 |
| Rubber products | 1.2 | 1.2 | 1.3 | 1.0 | 1.5 | 4.4 | 0.7 | 0.6 | 0.7 | 0.7 | 0.8 | 1.7 | 1.3 | 0.0 |
| Ceramics | 3.9 | 4.2 | 4.6 | 4.5 | 4.9 | 3.8 | 4.0 | 2.6 | 2.2 | 7.1 | 6.4 | 3.2 | 7.4 | 2.7 |
| Iron and steel | 10.9 | 4.6 | 4.0 | 2.0 | 8.3 | 4.7 | 5.0 | 6.6 | 3.9 | 8.1 | 3.4 | 4.6 | 14.0 | 0.6 |
| Non-ferrous metals | 3.0 | 3.1 | 3.3 | 2.6 | 4.4 | 6.3 | 3.7 | 1.5 | 3.5 | 9.9 | 4.2 | 1.5 | 5.0 | 2.3 |
| Metal products | 3.8 | 5.1 | 6.4 | 7.9 | 4.0 | 1.3 | 5.8 | 4.2 | 7.9 | 17.7 | 2.0 | 3.0 | 0.6 | 1.2 |
| General machinery | 12.2 | 15.0 | 20.4 | 19.3 | 26.0 | 2.8 | 7.1 | 5.4 | 7.0 | 4.9 | 9.7 | 16.5 | 7.9 | 6.2 |
| Electric machinery | 18.3 | 18.9 | 17.1 | 16.2 | 17.8 | 24.7 | 12.5 | 8.7 | 17.9 | 7.4 | 12.6 | 23.8 | 20.2 | 45.4 |
| Transport machinery | 8.6 | 8.3 | 5.7 | 6.5 | 5.0 | 0.3 | 15.6 | 7.6 | 30.9 | 9.5 | 11.1 | 5.0 | 15.9 | 1.3 |
| Precision instruments | 2.6 | 2.8 | 2.1 | 1.7 | 3.2 | 1.6 | 2.4 | 6.2 | 1.4 | 0.0 | 0.1 | 5.7 | 1.9 | 1.3 |
| Other manufacturing | 1.5 | 1.5 | 1.2 | 1.4 | 0.8 | 1.3 | 1.2 | 1.6 | 1.6 | 2.1 | 0.5 | 2.2 | 0.5 | 4.2 |

Source: Management and Coordination Agency, Kagakugijutu Kenkyu Chosa (Survey on Science and Technology Research), various issues.

Urata (1996) examines the patterns of intra-firm technology transfer by 126 Japanese firms in general machinery, electrical machinery, transport machinery and precision instruments to their affiliates in East Asia and identifies the factors that affect the outcome of technology transfer. According to his analysis, a large part of production technologies have been transferred to

*Table 6.6* Technology licensing by Japanese FDI firms (number of cases and percentage shares), 1986 and 1992

| 1986 Sector | To all firms | | | | | To overseas affiliates | | | | |
|---|---|---|---|---|---|---|---|---|---|---|
| | Asia | L. America | N. America | Europe | World | Asia | L. America | N. America | Europe | World |
| Total industry | 975 | 130 | 357 | 441 | 2,076 | 350 | 34 | 109 | 61 | 584 |
| Manufacturing | 930 | 128 | 340 | 426 | 1,992 | 336 | 33 | 102 | 55 | 553 |
| Food | 4 | 0 | 0 | 0 | 4 | 1 | 0 | 0 | 0 | 1 |
| Textiles | 109 | 1 | 8 | 9 | 127 | 74 | 0 | 0 | 0 | 74 |
| Wood and pulp | 3 | 1 | 22 | 16 | 44 | 0 | 0 | 0 | 0 | 0 |
| Chemical products | 99 | 27 | 57 | 108 | 325 | 26 | 2 | 2 | 10 | 40 |
| Iron and steel | 22 | 9 | 15 | 29 | 83 | 5 | 2 | 2 | 0 | 9 |
| Non-ferrous metals | 35 | 2 | 12 | 12 | 64 | 8 | 1 | 7 | 1 | 18 |
| General machinery | 71 | 5 | 19 | 78 | 182 | 20 | 0 | 9 | 4 | 35 |
| Electric machinery | 258 | 34 | 104 | 82 | 511 | 100 | 19 | 46 | 20 | 198 |
| Transport machinery | 196 | 36 | 42 | 55 | 385 | 61 | 3 | 9 | 4 | 80 |
| Precision instruments | 14 | 2 | 13 | 9 | 43 | 7 | 0 | 8 | 7 | 23 |
| Petro and coal products | 1 | 0 | 0 | 0 | 1 | 0 | 0 | 0 | 0 | 0 |
| Other manufacturing | 118 | 11 | 48 | 28 | 223 | 34 | 6 | 19 | 9 | 75 |

| 1992 | | | | | | | | | | |
|---|---|---|---|---|---|---|---|---|---|---|
| Total industry | 1,918 | 142 | 654 | 734 | 3,608 | 888 | 36 | 224 | 151 | 1,334 |
| Manufacturing | 1,843 | 138 | 619 | 714 | 3,442 | 865 | 36 | 208 | 145 | 1,287 |
| Food | 117 | 41 | 7 | 67 | 233 | 14 | 0 | 1 | 5 | 21 |
| Textiles | 51 | 5 | 18 | 33 | 110 | 27 | 2 | 4 | 4 | 38 |
| Wood and pulp | 1 | 0 | 4 | 6 | 11 | 0 | 0 | 1 | 1 | 2 |
| Chemical products | 197 | 15 | 130 | 245 | 606 | 55 | 6 | 18 | 15 | 96 |
| Iron and steel | 40 | 3 | 21 | 39 | 106 | 4 | 0 | 7 | 2 | 14 |
| Non-ferrous metals | 98 | 3 | 64 | 32 | 204 | 42 | 0 | 23 | 6 | 71 |
| General machinery | 109 | 5 | 35 | 31 | 183 | 37 | 3 | 19 | 19 | 78 |
| Electric machinery | 651 | 20 | 157 | 105 | 949 | 464 | 13 | 41 | 34 | 558 |
| Transport machinery | 401 | 34 | 103 | 108 | 709 | 143 | 8 | 58 | 33 | 263 |
| Precision instruments | 14 | 1 | 7 | 11 | 34 | 6 | 1 | 5 | 7 | 19 |
| Petro and coal products | 7 | 1 | 8 | 4 | 21 | 0 | 0 | 3 | 1 | 4 |
| Other manufacturing | 157 | 10 | 65 | 33 | 276 | 73 | 3 | 28 | 18 | 123 |

*Table* 6.6 (Continued)
Share of Intra-firm technology licensing in total technology licensing (%)

| Sector | 1986 | | | | | 1992 | | | | |
|---|---|---|---|---|---|---|---|---|---|---|
| | *Asia* | *L. America* | *N. America* | *Europe* | *World* | *Asia* | *L. America* | *N. America* | *Europe* | *World* |
| Total industry | 35.9 | 26.2 | 30.5 | 13.8 | 28.1 | 46.3 | 25.4 | 34.3 | 20.6 | 37.0 |
| Manufacturing | 36.1 | 25.8 | 30.0 | 12.9 | 27.8 | 46.9 | 26.1 | 33.6 | 20.3 | 37.4 |
| Food | 25.0 | – | – | – | 25.0 | 12.0 | 0.0 | 14.3 | 7.5 | 9.0 |
| Textiles | 67.9 | 0.0 | 0.0 | 0.0 | 58.3 | 52.9 | 40.0 | 22.2 | 12.1 | 34.5 |
| Wood and pulp | 0.0 | 0.0 | 0.0 | 0.0 | 0.0 | 0.0 | – | 25.0 | 16.7 | 18.2 |
| Chemical products | 26.3 | 7.4 | 3.5 | 9.3 | 12.3 | 27.9 | 40.0 | 13.8 | 6.1 | 15.8 |
| Iron and steel | 22.7 | 22.2 | 13.3 | 0.0 | 10.8 | 10.0 | 0.0 | 33.3 | 5.1 | 13.2 |
| Non-ferrous metals | 22.9 | 50.0 | 58.3 | 8.3 | 28.1 | 42.9 | 0.0 | 35.9 | 18.8 | 34.8 |
| General machinery | 28.2 | 0.0 | 47.4 | 5.1 | 19.2 | 33.9 | 60.0 | 54.3 | 61.3 | 42.6 |
| Electric machinery | 38.8 | 55.9 | 44.2 | 24.4 | 38.7 | 71.3 | 65.0 | 26.1 | 32.4 | 58.8 |
| Transport machinery | 31.1 | 8.3 | 21.4 | 7.3 | 20.8 | 35.7 | 23.5 | 56.3 | 30.6 | 37.1 |
| Precision instruments | 50.0 | 0.0 | 61.5 | 77.8 | 53.5 | 42.9 | 100.0 | 71.4 | 63.6 | 55.9 |
| Petro and coal products | 0.0 | – | – | – | 0.0 | 0.0 | 0.0 | 37.5 | 25.0 | 19.0 |
| Other manufacturing | 28.8 | 54.5 | 39.6 | 32.1 | 33.6 | 46.5 | 30.0 | 43.1 | 54.5 | 44.6 |

*Source*: Japan MITI, *Kaigai Toshi Tokei Soran* (Comprehensive Statistics of Foreign Investment), nos 3 1991 and 5 1994.

the Asian affiliates of Japanese firms, but transfer of sophisticated technologies such as design technology and development of new products has not been performed to a large degree. Among various management practices, job rotation and labour-management conferences are shown to be effective means of transferring technologies. Use of manuals in the local language, on-job training, small group activities and workshops in the local areas are also proven to be effective.

One of the most important contributions that the recipient countries of FDI expect from receiving FDI is technology transfer. In addition to formal technology licensing, which was discussed above, the recipient countries may acquire technologies through various channels such as accumulated work experiences at foreign firms and the spill over effect from the operations of foreign firms. The former type of technology transfer may be characterized as direct technology transfer, while the latter may be characterized as indirect technology transfer.

Both direct and indirect types of technology transfer have been detected by Urata and Iriyama (1997) in their study of the activities of foreign firms in different provinces in China. They found that the level of total factor productivity is significantly higher for foreign firms in comparison with local firms in most provinces, indicating the presence of direct technology transfer. Furthermore, they found that the level of total factor productivity of local firms is high in the provinces where the economic activities of foreign firms are significant. This finding may be interpreted as meaning that technologies are transferred from foreign firms to local firms indirectly.

## Japanese FDI and exports from host countries

Since the mid-1980s, foreign firms have played an important role in expanding exports from Asia, as foreign firms actively invested in Southeast and East Asia to expand their exports. As a consequence, FDI in these countries in most parts has been complementary to exports from the host countries.[9] This section examines the role of Asian affiliates of Japanese firms in expanding exports from Asia.

### Strong export orientation of Japanese firms in Asia

One notable characteristic of the sales behaviour of Asian affiliates of Japanese firms is their strong export orientation. This point can be seen clearly when compared to the behaviour of their affiliates in Latin America (Table 6.7). The share of exports in total sales (the export share, hereafter) for the Asian affiliates in manufacturing in 1992 was 33.9 per cent, while the corresponding share for the affiliates in Latin America was significantly smaller at 22.0 per cent. It should be noted that the export share for the Asian affiliates declined between 1986 and 1992, partly because the local demand increased rapidly on the back of rapid economic growth in Asia, and partly because the competitiveness of Asia's exports declined due largely to rising costs resulting from wage increases and currency appreciation.

The average export share for the manufacturing sector masks the wide variations observed for different subsectors.[10] The subsectors with strong export orientation are precision instruments, electrical machinery, food, wood and pulp, and general machinery, while those with weak export orientation are iron and steel, and transport machinery. The subsectors with strong export orientation are composed of two types of industries. One is natural resource intensive industries such as food, and wood and pulp. The other is parts and components intensive sectors (the industries that require a large number of parts and components for their production) such as precision instruments, electrical machinery and general machinery.

Strong export orientation of natural resource intensive sectors reflects the fact that Japanese firms have undertaken FDI in order to export and supply natural resources to Japan, which is poorly endowed with natural resources. Strong export orientation observed for the parts-intensive sectors may be explained by the strategy of Japanese firms, which was discussed in the third section pp. 153–8.

Japanese firms in parts-intensive sectors have adopted the 'inter-process specialization' strategy, or multinational and vertical integration strategy. Under the inter-process specialization strategy Japanese firms seek to minimize production cost by dividing the entire production process into a number of subprocesses and locating each subprocess in a country where that particular subprocess may be performed most efficiently, or at the least cost. For example, the production of TVs takes place in the following manner: TV tubes and integrated circuits (ICs) are produced at a parent firm in Japan and shipped to the Asian affiliate, where labour cost is low. There TVs are assembled by local labour using the imported parts, and a large part of the TVs are exported. The high share of intra-firm exports in total exports in these sectors shown in Table 6.8 appears to indicate that inter-process specialization strategy involving the parent companies and their foreign affiliates in Asia is actively pursued.

The low export orientation for transport machinery is due to the import protection policy applied to that subsector. Mainly because of the expectation that the development of an automobile industry leads to further industrialization by creating intensive and extensive backward linkages, the automobile industry has been the target of protection and promotion in a number of countries. Faced with import barriers, foreign firms have undertaken FDI to sell their products in the protected markets.

### Active intra-firm exports

The importance of intra-firm exports is observed for the exports of Asian affiliates, in particular regarding their exports to Japan. Indeed, the share of intra-firm exports in overall manufacturing exports was as high as 84.2 per cent in 1992 (Table 6.8). Although the shares are lower for their exports to other destinations, 50 per cent of Asian affiliates' exports to Asia, North America and Europe take the form of intra-firm trade. As noted above, the importance of intra-firm exports is especially significant for the machinery sectors, where inter-process specialization strategy has been pursued. One may interpret the high intra-firm export ratios as evidence that Japanese FDI is contributing to the expansion of Asian exports through the utilization of extensive market networks of Japanese firms, and thereby contributing to their economic growth.

One may also interpret the high intra-firm export ratio of Japanese firms as evidence of their exclusionary trading practices. As such, the benefits of FDI by Japanese firms are only captured by Japanese firms, and not shared by the host economies. One may argue, therefore, that the benefits of the recipient or host countries of Japanese firms would be greater if local firms were integrated into the production networks of Japanese firms. Although the underdevelopment of parts and components suppliers, or supporting industries, is a major obstacle to the extension of production networks involving local firms, there is a clear trend that local firms are becoming increasingly involved in the production networks of Japanese firms. For example, the local procurement ratio, measured as the ratio of the value of local procurement to the value of total procurement of the Asian affiliates of Japanese manufacturing firms increased from 42.2 per cent in 1986 to 48.5 per cent in 1992.

Table 6.7 Sales of foreign affiliates of Japanese firms in Asia and Latin America, 1986 and 1992 (percentage of total sales)

| Asia | 1986 | | | | | | | 1992 | | | | | | |
|---|---|---|---|---|---|---|---|---|---|---|---|---|---|---|
| | Local sales | Exports to | | | | | | Local sales | Exports to | | | | | |
| | | Total | Japan | Asia | N.America | Europe | L.America | | Total | Japan | Asia | N.America | Europe | L.America |
| Manufacturing | 54.7 | 45.3 | 15.8 | 12.8 | 10.2 | 4.6 | 0.3 | 66.1 | 33.9 | 15.8 | 11.2 | 3.7 | 2.0 | 0.1 |
| Food | 16.0 | 84.0 | 31.6 | 22.8 | 2.6 | 6.8 | 0.0 | 46.0 | 54.0 | 26.5 | 4.9 | 3.1 | 2.0 | 0.0 |
| Textiles | 45.7 | 54.3 | 10.3 | 24.2 | 8.5 | 4.3 | 0.0 | 56.1 | 43.9 | 14.2 | 12.3 | 7.2 | 6.6 | 0.2 |
| Wood and pulp | 44.5 | 55.5 | 25.5 | 12.0 | 14.6 | 3.3 | 0.0 | 50.2 | 49.8 | 47.2 | 0.3 | 2.4 | 0.0 | 0.0 |
| Chemical products | 78.5 | 21.5 | 3.8 | 16.8 | 0.5 | 0.2 | 0.0 | 64.7 | 35.3 | 4.9 | 28.6 | 0.4 | 0.2 | 0.0 |
| Iron and steel | 92.9 | 7.1 | 5.2 | 1.3 | 0.3 | 0.0 | 0.0 | 85.5 | 14.5 | 2.1 | 8.5 | 2.9 | 0.9 | 0.0 |
| Non-ferrous metals | 31.3 | 68.7 | 31.8 | 36.9 | 0.0 | 0.0 | 0.0 | 63.3 | 36.7 | 21.4 | 14.9 | 0.1 | 0.0 | 0.0 |
| General machinery | 44.5 | 55.5 | 31.4 | 4.9 | 13.5 | 4.6 | 0.0 | 53.0 | 47.0 | 23.6 | 11.3 | 2.1 | 9.8 | 0.0 |
| Electric machinery | 42.5 | 57.5 | 22.2 | 15.2 | 11.7 | 7.0 | 0.3 | 45.7 | 54.3 | 27.2 | 19.0 | 5.3 | 2.2 | 0.1 |
| Transport machinery | 74.4 | 25.6 | 5.3 | 1.6 | 13.6 | 2.6 | 0.8 | 92.6 | 7.4 | 1.7 | 1.0 | 3.9 | 0.5 | 0.0 |
| Precision instruments | 48.2 | 51.8 | 21.9 | 19.5 | 4.1 | 5.9 | 0.0 | 36.9 | 63.1 | 51.8 | 1.9 | 5.2 | 3.8 | 0.0 |
| Petro and coal products | 100.0 | 0.0 | 0.0 | 0.0 | 0.0 | 0.0 | 0.0 | 55.9 | 44.1 | 0.0 | 0.2 | 43.9 | 0.0 | 0.0 |
| Other manufacturing | 77.1 | 22.9 | 7.7 | 4.1 | 9.7 | 0.5 | 0.0 | 78.6 | 21.4 | 9.4 | 5.6 | 2.6 | 2.8 | 0.1 |

Table 6.7  (Contd.)

| Latin America | 1986 | | | | | | | 1992 | | | | | | |
|---|---|---|---|---|---|---|---|---|---|---|---|---|---|---|
| | Local sales | Exports to | | | | | | Local sales | Exports to | | | | | |
| | | Total | Japan | Asia | N.America | Europe | L.America | | Total | Japan | Asia | N.America | Europe | L.America |
| Manufacturing | 80.5 | 19.5 | 4.1 | 2.7 | 5.0 | 5.5 | 1.8 | 78.0 | 22.0 | 7.0 | 1.3 | 6.9 | 3.6 | 2.8 |
| Food | 44.8 | 55.2 | 18.2 | 0.0 | 11.1 | 23.5 | 0.1 | 27.8 | 72.2 | 15.4 | 8.9 | 14.9 | 24.9 | 5.7 |
| Textiles | 73.1 | 26.9 | 4.5 | 3.6 | 4.8 | 13.3 | 0.3 | 67.8 | 32.2 | 3.0 | 9.8 | 3.5 | 10.5 | 4.3 |
| Wood and pulp | 41.7 | 58.3 | 31.3 | 0.0 | 14.3 | 6.0 | 6.7 | 21.1 | 78.9 | 35.0 | 1.3 | 28.9 | 12.5 | 1.4 |
| Chemical products | 93.6 | 6.4 | 0.0 | 4.7 | 0.0 | 0.0 | 1.7 | 86.0 | 14.0 | 0.5 | 2.0 | 4.0 | 4.1 | 2.8 |
| Iron and steel | 82.4 | 17.6 | 2.9 | 6.6 | 4.7 | 1.3 | 1.4 | 100.0 | 0.0 | 0.0 | 0.0 | 0.0 | 0.0 | 0.0 |
| Non-ferrous metals | 82.7 | 17.3 | 0.0 | 11.6 | 0.0 | 0.0 | 5.7 | 84.3 | 15.7 | 10.9 | 0.0 | 0.0 | 0.0 | 0.0 |
| General machinery | 97.4 | 2.6 | 0.0 | 0.3 | 0.0 | 0.0 | 1.4 | 74.4 | 25.6 | 0.1 | 0.0 | 0.3 | 0.0 | 25.2 |
| Electric machinery | 94.5 | 5.5 | 0.8 | 0.0 | 2.5 | 1.9 | 0.3 | 96.3 | 3.7 | 0.2 | 0.1 | 1.6 | 0.4 | 1.4 |
| Transport machinery | 82.0 | 18.0 | 0.0 | 0.0 | 0.5 | 16.6 | 0.8 | 80.4 | 19.6 | 12.0 | 0.1 | 5.9 | 0.0 | 1.4 |
| Precision instruments | 65.4 | 34.6 | 1.2 | 0.0 | 1.1 | 6.2 | 26.1 | 100.0 | 0.0 | 0.0 | 0.0 | 0.0 | 0.0 | 0.0 |
| Petro and coal products | 0.0 | 0.0 | 0.0 | 0.0 | 0.0 | 0.0 | 0.0 | 99.6 | 0.4 | 0.0 | 0.0 | 0.0 | 0.0 | 0.4 |
| Other manufacturing | 76.0 | 24.0 | 1.0 | 0.4 | 17.3 | 0.9 | 4.3 | 61.2 | 38.8 | 3.0 | 0.0 | 35.8 | 0.0 | 0.1 |

Source: Japan, MITI, Kaigai Toshi Tokei Soran (Comprehensive Statistics on Foreign Investment by Japanese Firms), vols 3 and 5, 1989 and 1994.

*Table 6.8*    Intra-firm trade by overseas affiliates in Asia of Japanese firms, 1986 and 1992 (percentage share of overall trade)

| | 1986 | | | | | 1992 | | | | |
|---|---|---|---|---|---|---|---|---|---|---|
| | Local sales | Exports to | | | | Local sales | Exports to | | | |
| | | Japan | Asia | N. America | Europe | | Japan | Asia | N. America | Europe |
| Manufacturing | 8.9 | 76.5 | 20.1 | 58.1 | 50.5 | 6.3 | 84.2 | 44.4 | 62.4 | 47.6 |
| Food | 0.0 | 87.0 | 0.0 | 0.0 | 0.0 | 7.6 | 85.4 | 26.3 | 51.9 | 50.1 |
| Textiles | 8.0 | 57.7 | 0.7 | 49.1 | 31.9 | 4.3 | 36.1 | 23.0 | 1.1 | 0.9 |
| Wood and pulp | 0.0 | 27.7 | 0.0 | 0.0 | 0.0 | 0.0 | 57.9 | 0.0 | 0.0 | 0.0 |
| Chemical products | 2.6 | 83.9 | 1.4 | 0.0 | 0.0 | 2.4 | 49.0 | 3.2 | 11.5 | 17.6 |
| Iron and steel | 3.2 | 100 | 0.0 | 0.0 | 0.0 | 0.0 | 29.0 | 0.0 | 23.3 | 0.0 |
| Non-ferrous metals | 15.1 | 99.2 | 0.0 | 0.0 | 0.0 | 0.8 | 82.6 | 55.1 | 0.0 | 0.0 |
| General machinery | 29.9 | 94.7 | 45.5 | 46.9 | 58.8 | 3.0 | 96.7 | 55.6 | 54.3 | 93.9 |
| Electric machinery | 9.6 | 73.0 | 26.4 | 60.6 | 52.0 | 8.0 | 90.0 | 53.7 | 82.6 | 58.0 |
| Transport machinery | 9.1 | 46.0 | 0.0 | 84.5 | 29.4 | 7.2 | 73.9 | 57.9 | 71.2 | 28.3 |
| Precision instruments | 59.6 | 86.1 | 46.6 | 74.1 | 92.4 | 32.4 | 96.5 | 77.9 | 51.1 | 50.8 |
| Petro and coal products | 0.0 | 0.0 | 0.0 | 0.0 | 0.0 | 0.0 | 0.0 | 0.0 | 0.0 | 0.0 |
| Other manufacturing | 0.0 | 88.5 | 7.4 | 16.7 | 47.2 | 6.3 | 67.0 | 49.8 | 25.3 | 18.4 |

*Source*: Japan, MITI, *Kaigai Toshi Tokei Soran* [Comprehensive Statistics on Foreign Investment by Japanese Firms], vols 3 and 5, 1989 and 1994.

## *The contribution of Japanese FDI for export expansion of host countries*

It is of interest to examine the magnitude of the contribution of the Asian affiliates of Japanese firms towards the promotion of exports of the host countries. A lack of detailed information on the exports of the Asian affiliates of Japanese firms makes it difficult to measure the exact size of the contribution. According to the Results of the Basic Survey of Business Structure and Activity (MITI), the exports of the affiliates of Japanese firms in East Asia under the survey amounted to $16.7 billion in 1991. Since the total merchandise exports of the East Asian countries for the same year was $392 billion, the share or contribution of the Asian affiliates of Japanese firms for the exports of the host countries was 4.4 per cent (Table 6.9).

There are wide variations in the shares of the affiliates of Japanese firms in the host countries' exports among the Asian countries. Japanese firms contributed significantly to the promotion of exports for Singapore and Malaysia, both in terms of absolute values and in relation to the overall exports of these countries. The affiliates of Japanese firms in Singapore and in Malaysia respectively exported manufactured goods worth $6.2 and $5.5 billion in 1991, amounting to 10.7 and 9.6 per cent of total merchandise exports of the respective countries.

By contrast, the contribution of the affiliates of Japanese firms was small in China and Indonesia, since their respective shares in total exports of these countries were less than 1 per cent.[11]

169

*Table 6.9*    Exports of Asian affiliates of Japanese firms, 1991

| Country | Manufactures exports by Japanese firms ($ million) (1) | Merchandise exports ($million) (2) | Share of exports by Japanese firms in total exports (1)/(2) (%) |
|---|---|---|---|
| NIEs | 11,021 | 233,122 | 4.73 |
| Hong Kong | 1,970 | 29,693 | 6.63 |
| Korea | 912 | 69,582 | 1.31 |
| Singapore | 6,243 | 58,312 | 10.71 |
| Taiwan | 1,896 | 75,535 | 2.51 |
| ASEAN4 | 5,522 | 100,339 | 5.50 |
| Indonesia | 171 | 29,635 | 0.58 |
| Malaysia | 3,210 | 33,533 | 9.57 |
| Phlippines | 397 | 8,840 | 4.50 |
| Thailand | 1,744 | 28,331 | 6.16 |
| China | 137 | 58,919 | 0.23 |

*Sources*: Japan, MITI, Results of the Basic Survey of Business Structure and Activity: 1992, vol. 3; 1994. ADB, Key Indicators of Developing Asian and Pacific Countries: 1995.
*Notes*: Comparison of columns (1) and (2) is not exact at least for the following three reasons: (1) refers to manufactures exports while (2) refers to merchandise exports; (1) refers to Japanese fiscal year, April–March, while (2) refers to the calendar year; (1) only considers exports by Asian affiliates of Japanese firms with paid-in capital exceeding $million, while (2) considers all exports.

## Conclusions

Foreign direct investment has increased its importance as a mechanism firms have for transferring technologies and establishing marketing, and procuring networks for efficient production and sales internationally. Through FDI, foreign investors benefit from utilizing their assets and resources efficiently, while FDI recipients benefit from acquiring technologies and from getting involved in international production and trade networks. Indeed, rapid economic growth in East Asia in recent years is to some extent attributable to active FDI undertaken in the region.

After a significant decline in the early 1990s, Japanese FDI started to increase again in 1993 and has continued its expansion mainly because of the appreciation of the Japanese yen. Although the pace of the increase in Japanese FDI has slowed down with the depreciation of the yen in recent years, there appears to be a consensus among researchers that Japanese FDI as well as world FDI is going to continue its expansion in the future. Indeed, the overseas production ratio, which is defined as the proportion of overseas production in total (local and overseas) production, for Japanese firms, is predicted to increase from 10.0 per cent in 1995 to 18.6 per cent in 2010 through FDI expansion (Economic Planning Agency, 1997).

At least two factors are behind such predictions. First, recognizing the importance of attracting FDI for economic growth, countries are likely to adopt policies to promote FDI inflow. Second, multinational firms expand FDI, in order to achieve most efficient production on a global scale. Specifically, multinational firms locate a production process in a country where that particular process can be carried out at the least cost. This type of overseas production strategy is actively undertaken by the machinery sectors which use a large number of components in their production. Furthermore, overseas production for Japanese firms is likely to expand, as the factors of production, namely labour and capital, in Japan are expected to decline in the near future, as a result of the ageing and the absolute decline of the Japanese population.

This chapter has found a number of factors in developing countries that have attracted Japanese FDI. High economic growth, abundance of low wage workers, trade and FDI liberalization and pro-FDI policies are among them. Additional factors that would attract FDI may be identified from the problems faced by Japanese firms in their businesses overseas. Several questionnaire surveys conducted by the Japanese government institutions and private research institutes including MITI, JETRO and the Export-Import Bank of Japan, found that Japanese firms are faced with various problems in their business in developing countries.[12] The shortage of skilled workers, the underdevelopment of supporting industries and insufficient infrastructure are the three problems most often pointed out by Japanese firms. These findings indicate that Japanese firms would be attracted to a country where skilled workers, supporting industries and infrastructure are available.

So far Japanese FDI in developing countries has been concentrated in the NIEs, ASEAN countries and China in East Asia. But as was noted earlier, other countries in Asia such as India and Vietnam started to attract Japanese FDI, as the attractiveness of the NIEs, ASEAN countries and China declined with the emergence of internal problems such as the shortage of skilled labour, lack of infrastructure and environmental problems. This new development indicates that developing countries which can offer a desirable environment for Japanese firms can attract Japanese FDI.

The governments in developing countries interested in attracting FDI to promote economic growth should pursue three kinds of policies, which are related to each other. One type of policy is sound macro-economic policy, which would lead to low inflation and economic growth. Another is micro-economic policy such as trade and FDI liberalization which would reduce the distortions in the market to increase competition and to realize efficient resource allocation. A reliable legal system that supports the activities of foreign firms is one of the most important micro-economic policies. The third type of policy is to provide infrastructure (broadly defined) such as transport facilities and education that are necessary for promoting economic development. In many cases, developing countries face a shortage of resources, including foreign exchange and human resources, to carry out these policies. To deal with these problems, they should utilize various types of support from foreign countries through bilateral and multilateral arrangements.

## Acknowledgements

The author wishes to thank Akie Iriyama for his research assistance and Dr Nagesh Kumar and other participants for helpful comments and discussions.

## Notes

1 This section extends Urata (1993a) and Kawai and Urata (1995).
2 This observation applies to new FDI. A lack of data precludes one from obtaining accurate statistics, but there are signs that Japanese FDI undertaken with reinvested earnings has continued to grow in the ASEAN countries. See, for example, Okamoto and Urata (1994) for the case of Malaysia.
3 See Urata (1993b) for a detailed discussion.
4 *Kaigai Toshi Kenkyusho Ho* (Journal of Research Institute for International Investment and Development) 23 (1), January 1997, The Export-Import Bank of Japan.
5 The MITI conducts a comprehensive survey of overseas activities of Japanese firms every three years, starting in 1980. The results of the survey are published under the title of *Kaigai Toshi Tokei Soran* (Comprehensive Survey of Overseas Activities of Japanese Firms). In the 1992 survey, a questionnaire was sent to 3,378 Japanese multinationals, 1,594 of whom responded. The respondents covered the activities of 7,108 overseas affiliates. The MITI also conducts an annual survey with a more limited number of questions except for the years of the Comprehensive Survey. The results of the survey are published under the title of *Wagakuni Kigyo no Kaigai Jigyo Katsudo* (Overseas Activities of Japanese Firms). Motives for FDI are asked for in the comprehensive surveys but not in the annual surveys.
6 Kumar (1995) presents theoretical explanations and empirical findings concerning technology acquisition by developing countries.
7 Under OEM contracts, contractee firms make the entire product for the contractor according to the designs and specifications (and brand name) of the contractor. Subcontracting refers to such practice under which only part of production such as assembling is undertaken by a subcontractee.
8 For example, see Urata (1990) and Kim and Lee (1990) respectively on the experiences in Japan and in Korea in the post-World War II period.
9 Kawai and Urata (1995) find that Japanese FDIs are complements to Japan's exports as well as imports by analysing their relationships statistically in the framework of the gravity model. Petri (1995) obtains similar results from his analysis of FDI and international trade of developing countries.
10 Kumar (1996a) examines the export orientation of Japanese firms in developing countries.
11 The contribution of all foreign firms, including Japanese firms, to the export expansion of the host countries is difficult to measure, as necessary information is not readily available for most countries. One exception is China, where the share of total exports conducted by foreign firms is officially reported. According to the figures released in the Chinese *Statistical Yearbook*, the share increased rapidly from 16.8 per cent in 1991 to 31.5 per cent in 1995.

12 See the Comprehensive Statistics of Overseas Activities of Japanese Firms (MITI), The White Paper on Foreign Direct Investment (JETRO) and Annual Survey of the Outlook of Japanese Foreign Direct Investment (Export-Import Bank of Japan).

# Part III

# EMERGING SOURCES OF FDI AND TECHNOLOGY

This part covers trends in the emerging sources of FDI and technology, their relative importance for the developing countries as sources of FDI and the prospects for developing countries. It examines the trends and patterns in FDI outflows from the newly industrializing economies, especially those in Asia which have become significant sources of FDI in the past ten years. The emerging multinationals from developing countries have become important sources of FDI inflows for other developing countries and their importance as sources of FDI is particularly striking for ASEAN and some Latin American countries. The implications of this trend for developing countries as potential hosts of FDI are also examined.

# 7

# EMERGING OUTWARD FOREIGN DIRECT INVESTMENTS FROM ASIAN DEVELOPING COUNTRIES

## Prospects and implications

*Nagesh Kumar*

## Introduction

FDI outflows have largely originated in the traditionally capital surplus industrialized countries. Hence, the emergence of FDI outflows from some developing countries, first noticed in the early 1970s (see Lecraw, 1977), has aroused considerable academic and policy interest. After a decade of moderate but steady growth, annual magnitudes of outflows of FDI from developing countries have grown at a dramatic pace since the mid-1980s and have become a significant proportion of global FDI outflows. We argue that this sudden spurt in the magnitudes of annual flows of FDI from developing countries coincided with a change in the motivation for these flows. In the earlier round, developing country FDIs were of essentially market defensive type and were generally made in the neighbouring countries at a lower step in terms of levels of economic and technological development. In the current round, enterprises from developing countries, especially newly industrializing economies (NIEs) have increasingly used outward FDI as a strategic tool to strengthen their competitiveness in major markets. The evidence offered in support of this assertion consists of trends and patterns in the FDI outflows from developing countries indicating their motive for supporting price and non-price competitiveness of their manufactured exports, and case studies of government policy changes favouring outward flows in selected important sources of FDI among Asian developing countries. The chapter also comments on the implications of these trends for developing countries which are striving to attract greater magnitudes of FDI inflows as a bundle of much needed technology, capital and entrepreneurship in their effort to industrialize.

The structure of the chapter is as follows. The next section summarizes the trends and patterns in outflows of FDI from developing countries. The third section discusses the ways in which developing country enterprises have used outward FDI as means of improving their non-price competitiveness; and the next section similarly deals with trends with respect to overseas investment for strengthening price competitiveness. The fifth section reviews the trends in government policy towards outward investment in a few important sources of FDI outflows in the developing world to assess the motivations for official promotion. The sixth section analyses the implications of the recent trends for FDI prospects for poorer developing countries, and finally, the chapter is concluded with a few policy remarks for developing countries.

## Evolution of outflows of FDI from developing countries

Table 7.1 summarizes the data on global outflows of FDI. It would appear from it that developing countries, which contributed just 2 per cent of global FDI outflows in the early 1980s, currently provide 15 per cent of global flows. The bulk of these flows originate in developing countries in the East, Southeast and South Asian countries which contributed $41.5 billion of $47 billion (or 88 per cent) of all FDI outflows originating in developing countries. To some extent the dramatic rise in the annual magnitudes of FDI outflows from developing countries in recent years is on account of a huge jump in annual outflows of FDI from Hong Kong since 1992. The annual outflows from Hong Kong increased by nearly nine times within a span of five years – 1991 to 1995. As a result, Hong Kong alone now accounts for 53 per cent of all FDI outflows from developing countries. To the extent that outflows originating in Hong Kong represent ethnic investments, partly round-tripping of Chinese capital, and partly FDI outflows from other sources rerouted through Hong Kong (see, for instance, Zhang and Van Den Bulcke, 1996) the recent rise in FDI outflows from developing countries may be overblown. Yet, Table 7.1 shows the emergence of countries like South Korea, Taiwan, Singapore, China and even Malaysia each providing about $3 billion in annual outflows, not an insubstantial amount considering the fact that inflows of FDI to all developing countries till the beginning of 1990s averaged around $30 billion a year.

The stock of outward FDI made by select Asian countries presented in Table 7.2 also underscores the observation made from Table 7.1 – that the growing importance of developing countries as sources of FDI is a relatively recent phenomenon, having accelerated over the past six or seven years. The outward stock of the select Asian countries has grown by 273 per cent over the five year period 1990–5 compared to 209 per cent for all outward FDI stocks originating from developing countries. Over the same period, the global outward stocks of FDI grew by only 62 per cent. The substantially higher rate of growth of developing country outflows has seen their share in total

*Table 7.1* FDI outflows originating in developing countries, 1982–95 ($ million)

| | 1982–7 Annual average | 1984–9 Annual average | 1990 | 1991 | 1992 | 1993 | 1994 | 1995[a] |
|---|---|---|---|---|---|---|---|---|
| Global outflows | 67,876 | 121,630 | 240,253 | 210,821 | 203,115 | 225,544 | 230,014 | 317,849 |
| Developing countries | 1,321 | 7,621 | 17,765 | 8,853 | 21,629 | 32,981 | 38,612 | 47,001 |
| (percentage share) | (2) | (6) | (7) | (4) | (11) | (15) | (17) | (15) |
| Selected Asian countries (total) | 812 | 5,147 | 12,276 | 8,651 | 17,379 | 29,263 | 33,003 | 41,527 |
| China | 333 | 581 | 830 | 913 | 4,000 | 4,400 | 2,000 | 3,467 |
| Hong Kong | | 1,833 | 2,448 | 2,825 | 8,254 | 17,713 | 21,437 | 25,000 |
| Korea, Republic of | 106 | 137 | 1,056 | 1,500 | 1,208 | 1,361 | 2,524 | 3,000 |
| Malaysia | | 233 | 532 | 389 | 514 | 1,325 | 1,817 | 2,575 |
| Singapore | 178 | 286 | 2,034 | 1,024 | 1,317 | 1,784 | 2,177 | 2,799 |
| Taiwan | 162 | 1,999 | 5,243 | 1,854 | 1,869 | 2,451 | 2,460 | 3,822 |
| Thailand | 29 | 41 | 140 | 167 | 147 | 221 | 493 | 904 |

*Note*:   a Provisional.
*Source*: Compiled from UNCTAD *World Investment Reports 1994* and *1996*.

*Table 7.2* Stock of outward foreign direct investments made by select Asian countries, 1980–95 ($ million)

| Country | 1980 | 1985 | 1990 | 1995 | Percentage change between 1990 and 1995 |
|---|---|---|---|---|---|
| South Korea | 142 | 487 | 2,172 | 1,1079 | 410 |
| Taiwan | 101 | 215 | 12,888 | 65,000[2] | 404 |
| Hong Kong | 1,800[1] | 9,441 | 18,930 | 85,156 | 350 |
| Singapore | 652 | 1,320 | 4,277 | 13,842 | 224 |
| China | 39 | 131 | 2,488 | 17,268 | 594 |
| Malaysia | 414 | 749 | 2,283 | 8,903 | 290 |
| Thailand | 13 | 14 | 398 | 2,333 | 486 |
| India | 149 | 180 | 290 | 850[3] | 193 |
| Indonesia | −1 | 49 | 25 | 110 | 340 |
| Total for the above countries | 1,510 | 12,586 | 43,751 | 203,691 | 365 |
| Total all developing countries | 6,167 | 21,222 | 69,369 | 252,453 | 264 |
| Percentage share of all countries | (1) | (3) | (4) | (9) | |
| All countries | 513,740 | 685,549 | 1,684,136 | 2,730,146 | 62 |

*Sources*: Compiled on the basis of UNCTAD, *World Investment Report 1996*: 1   Lall (1984);
2   Taiwan's Ministry of Economic Affairs as reproduced in *Financial Times*, 10 October 1996; and 3   own estimates.
*Note*: The totals for developing countries take into account the additional sources used for Taiwan and India and hence may be at variance with UNCTAD figures for 1995.

outward FDI stocks more than doubling from 4 per cent in 1990 to 9 per cent in 1995. The pre-eminent position of Hong Kong as a source of outward FDI among developing countries is underlined again. Taiwan also appears to have accounted for a considerable slice of outward FDI stock among developing countries. A few countries account for the bulk of FDI stocks originating in the developing world in value terms, as indeed is the case with OECD countries' outflows. However, the emergence on the scene of other source countries such as Singapore, Malaysia, Thailand, India and Indonesia is quite apparent considering the substantial growth in the outward stocks over the 1990–5 period.

## Outward FDI as a means of improving non-price competitiveness

International competitiveness is a combination of price and non-price competitiveness. The access to captive sources of information on markets and changing consumer tastes, the quality of after sales service provided and access to established brand names are, among others, important aspects of non-price competitiveness. It could be an important determinant of market access especially in the case of skill and knowledge intensive goods. Hirsch (1970) was among the first to recognize the importance of non-price competitiveness and emphasized the role of the 'proprietary associations abroad' in augmenting non-price competitiveness. In the past, developing country enterprises generally secured proprietary associations abroad by tying up with the conventional MNEs (i.e. those originating in the industrialised countries) in the form of subcontracting arrangements, strategic alliances or inward FDI (or controlling affiliations). In the recent period, developing country enterprises have also increasingly set up affiliates abroad that support their own export effort. In other words, outward FDI has increasingly been used as a means of securing proprietary associations abroad by developing country enterprises in recent times.

The outward investment to improve the non-price aspect of competitiveness of developing country goods has taken three forms. It has included: (1) trade supporting investments abroad; (2) strategic asset seeking acquisitions to complement their asset bundles; and (3) strategic access to markets seeking investments made in emerging trading blocs, to escape protectionist tendencies by becoming 'insiders'.

### Trade supporting investments

The trade supporting investments have generally been made in major industrialized countries and include establishment of affiliates to develop marketing networks in the host countries or regions and provide after sales services. In recent years, a substantial part of outward FDI of developing countries, especially in industrialized countries, has gone into trading. The sectoral break-down of South Korean FDI summarized in Table 7.3 shows, for instance, that nearly half of all Korean overseas investments in North America and Europe in terms of numbers and about 40 per cent in terms of value (as at the end of 1994) have been in trading activity. The respective comparable figures for global Korean investments are 19 per cent and nearly 23 per cent. This suggests that a much larger part of Korean FDI in industrialized countries has gone in creating trading infrastructure and support facilities. Furthermore, the investment in creating marketing infrastructure in major markets is a relatively recent development. In the earlier phase of their evolution, Korean enterprises obtained 'proprietary associations abroad' by their affiliations with the conventional MNEs in the form of subcontracting deals which involved selling under the brand names of foreign MNEs. Gradually, they are securing the associations with their own subsidiaries. It is apparent from the fact that nearly 90 per cent of all investments made in trading by Korean companies so far, were made in the first half of the 1990s (Bank of Korea, 1995). This has helped in bringing down the proportion of Korean goods sold abroad under the brand names of foreign firms over the years to about half (*International Herald Tribune,* 19 September 1995).

Even though Chinese enterprises have begun investing abroad relatively recently, it is already evident that outward FDI is increasingly seen by them as a means of promoting international competitiveness by increased foreign presence. A survey of Chinese enterprises with overseas investment in non-trade sectors in Shanghai, Beijing and Fujian provinces, conducted by the Institute of World Economy, Fudan University in 1989, indicated that opening up new markets, and acquisition of first hand information on markets were considered as motives for investing abroad by 94 per cent of enterprises. Furthermore, 83 per cent of enterprises felt that outward investment helped them in improving their customer relations and credibility in overseas markets (Gang, 1992). Similarly, overseas investments by Indian enterprises in

*Table 7.3* Korean overseas FDI stock by destination and industry, 1994 (number and US$ thousand)

| Sectors | | Destination | | | |
|---|---|---|---|---|---|
| | | North America | Europe | Southeast Asia | Total |
| All Industries | Number | 662 | 245 | 2,845 | 4,161 |
| | Value | 2,703,845 | 981,300 | 3,201,820 | 7,648,792 |
| Manufacturing | Number | 184 | 82 | 2,201 | 2,652 |
| | Value | 1,153,762 | 504,798 | 2,284,283 | 4,191,101 |
| Trading | Number | 315 | 130 | 312 | 807 |
| | | (47.58) | (53.06) | (10.97) | (19.39) |
| | Value | 1,062,321 | 408,306 | 235,782 | 1,750,562 |
| | | (39.29) | (41.61) | (7.36) | (22.89) |

*Source*: Extracted from Bank of Korea (1995).
*Note*: Figures in parentheses are percentages with respect to total for all industries.

the period since the mid-1980s were increasingly concentrated in industrialized countries bringing their share in outward stocks up from 11 per cent in 1980 to 19 per cent by 1993. Nearly 70 per cent of these investments in industrialized countries are in services, with trading alone accounting for a quarter. Another 18 per cent of total Indian investments in services in the industrialized countries is actually in engineering and consultancy services generally made by Indian software and engineering enterprises to secure consultancy contracts and to provide after sales services to their clients. The nature of these investments is quite close to those of trading investments (Kumar, 1996c).

### Strategic asset seeking investments

Access to established brand names and novel product technology constitutes an important aspect of non-price rivalry. A considerable proportion of developing countries' FDI in the recent period has gone into acquisitions of industrialized country enterprises in an attempt to augment the asset bundles of investing enterprises with complementary assets, most often established brand names. For instance, Samsung Electronics of Korea acquired AST Research (computers), USA; and LG Electronics acquired Zenith Electronics, the last TV producer in the United States, to get access to the brand names and marketing networks of the US companies. Daewoo Electronics was about to take over the consumer electronics part of Thomson SA of France and to become the largest TV manufacturer in the world, had the deal not been vetoed by the French government. Among the Taiwanese companies, Acer Computers acquired Counterpart Computers, USA; Umax Data Systems acquired a computers manufacturing unit of Radius Inc.; and United Microelectronics Corporation acquired a controlling interest in California based Catalyst Semiconductors Inc., for securing the technology, market access and brands of the US companies. Some Taiwanese enterprises have pursued an aggressive acquisition strategy in pursuit of brand names and market shares in the US market, e.g. President Enterprises Corporation which acquired Wyndham Foods Inc. in 1990, Famous Amos Chocolate Chip Cookie Corp. in 1992, and Mother's Cakes and Cookies in 1996 (*International Herald Tribune*, 13 February 1996). Also, Proton, a car manufacturer of Malaysia, took over Lotus, a British manufacturer of sports car to get access to the latter's engineering capabilities.

### Strategic access to markets seeking investments in trading blocs

A considerable proportion of overseas investments by developing country enterprises (or emerging MNEs) has also been made in greenfield projects to escape protectionist threats in the industrialized countries. For instance, most Korean electronics exports to Europe such as VCRs, colour TVs, CD players, video tapes and car stereos are currently subject to anti-dumping charges by the EU. Microwave ovens are subject to quotas. Furthermore, the formation of the Single European Market and NAFTA has brought a wide spread perception of threat of discrimination against extra-regional supplies. Exporters from Japan and the US have responded to the threat of protectionism arising from the Single European Market by investing within the EU and claiming access to it as an 'insider' (see Kumar, 1994b, among others, for illustrations). Korean chaebols, which dominate Korean FDI and large Taiwanese companies, have also responded in the same manner. As a result, Korean FDI into the EU has increased rapidly since 1990. Of the $1,091 million investment made in Europe by Korean companies by the end of 1994, nearly 90 per cent had been made in the last four years. Given the fact that the protectionist threat is greatest for consumer electronics products, the bulk of these investments have been made in consumer electronics assembly. All the leading Korean MNEs have invested in the EU. Samsung, for instance, invested in a $15 million TV plant at Billingham, UK, in 1990; a $30 million semiconductor assembly plant in Porto, Portugal, 1994; a $16 million excavator plant in Harrogate, UK, 1995; and a $700 million consumer electronics complex, Wynyard, UK, over the period 1994–2000; and a $120 million CPT plant, Berlin, Germany, over the period 1994–7. Finally, it decided to set up its European headquarters in London. LG (Lucky Goldstar) invested in a $10 million VCR plant in Worms, Germany in 1986, a $10 million refrigerator/freezer plant, Naples, Italy in 1990, and a $40 million colour TV and microwave oven plant in Newcastle, UK in 1994–5. Daewoo Group set up a $42 million VCR plant in Antrim, Northern Ireland in 1988, a $23 million microwave oven plant in Longwy, France in 1988, a $10 million excavator plant in Frameries, Belgium, 1990, a $37 million CTV plant and two CPT factories worth $150 and $138 million in France, during 1993–5. Hyundai is investing $3.7 billion in a microchips plant in Scotland (*Financial Times*, 10 February 1995; *International Herald Tribune*, 9 October 1996, among other sources).

Similarly, NAFTA has prompted Korean companies to set up greenfield plants within the customs unions to escape possible protectionism. Samsung has invested $212 million in a TV plant located in Tijuana, Mexico and has planned a $582 million expansion. Daewoo Corp. has invested $250 million in a TV picture tubes plant in Mexicali, Mexico. The bulk of the production of these plants is meant for the US, a partner of Mexico in NAFTA (*International Herald Tribune*, 24 May 1996). Major Taiwanese corporations, e.g. Acer, Tatung and Taco Electric, have also set up manufacturing bases in Europe and North America. Taiwan Semiconductor Manufacturing Co. set up a $1.2 billion microchip plant in the US (*International Herald Tribune*, 13 February 1996). Towards the end of the 1980s, Hong Kong enterprises were also increasingly responding to the rising protectionism in the industrialized countries and the formation of trade blocks. A World Bank Survey indicated that most Hong Kong investors were concerned about the 1992 deadline

*Table 7.4*  Industrialized countries' share in outward FDI stock of selected Asian developing countries, 1980 and 1993 (per cent)

| Home country | Industrialized countries' share in outward FDI stock | |
| --- | --- | --- |
| | 1980 | 1993 |
| China | 34 | 71 |
| Hong Kong | 8 | 18 |
| India | 11 | 19 |
| Singapore | 9 | 21 |
| South Korea | 32 | 48 |

*Sources*: Derived from UN-TCMD, 1993, Table II.3; Bank of Korea (1995), and Kumar (1996c).

for unification of the EU and had actively looked into investments within the Union with English speaking Ireland and UK as the first choices, and lower-labour-cost Spain as the second (World Bank, 1989).

To sum up, therefore, developing country enterprises have increasingly used outward FDI as a means of strengthening their non-price competitiveness in recent years by setting up trading subsidiaries, strategic asset seeking acquisitions and greenfield investments within the emerging trade blocs to gain insider status. A combination of these trends explains the rising concentration of outward FDI of developing countries in industrialized countries as revealed by Table 7.4 (Dunning also makes a similar observation in Chapter 3 of this volume).

## Outward FDI as a means of improving price competitiveness

Enterprises in the East Asian newly industrializing economies are also using outward FDI to improve the price competitiveness of their goods, which has suffered adversely from rising wages and currency appreciations over the past ten years, by moving their production to cheap labour locations. Sometimes these investments are also driven by the availability of preferential access to major markets, e.g. the Mediterranean, East or Central European, or to Lomé Convention (or ACP) countries that enjoy preferential access to the European Union market, by the loss of GSP preferences and exhaustion of MFA quotas in garments and textiles industry. This is illustrated below with examples from South Korea, Taiwan and Hong Kong.

### South Korea

The Korean won has tended to appreciate owing to rising current account surpluses in the late 1980s and early 1990s (Chaponniere, 1992). Korean international competitiveness was further affected by rising real wages. Between

1978 and 1987 alone, the South Korean production workers' hourly wages doubled (World Bank, 1989) and have risen by 16 per cent a year on average in nominal terms since then (*Economist*, 14 September 1996). Finally, Korean export competitiveness was also eroded by the loss of GSP preferences in the US market as South Korea was considered developed enough to no longer need trade preferences. The further expansion of South Korean exports of footwear and apparel etc. has been limited by the exhaustion of MFA quotas. Korean enterprises attempted to make up for these developments by relocating labour intensive production in Southeast and South Asian countries to take advantage of low cost labour and also of GSP preferences and MFA quotas. Table 7.5 shows that a substantial proportion of Korean manufacturing FDI in the Southeast Asian countries, and some South Asian countries such as Bangladesh and Sri Lanka, has been in export-oriented labour intensive industries such as textiles and garments and footwear and leather goods. That suggests that Korean enterprises have used these countries as export platforms for relocating these industries owing to the availability of cheap labour, GSP benefits and often also unfilled MFA quotas. Korean companies are also setting up plants in Eastern European and Mediterranean countries which combine the benefit of relatively cheaper wages with a preferential access to the EU markets. For instance, Samsung has invested in Hungary; Daewoo has invested in a $759 million car factory in Craiora, Romania and a TV and microwave oven plant in Poland (*Financial Times*, 10 February 1995, and

*Table 7.5* Sectoral distribution of overseas FDI stock of South Korea in selected Asian countries, end 1994 (US$ thousand)

| Host countries | Sectors | | | |
|---|---|---|---|---|
| | All industries | Manufacturing | Textiles and clothing | Leather and footwear |
| Philippines | 159,889 | 153,549 | 29,426 (19) | 9,843 (6) |
| China | 1,104,924 | 996,079 | 143,386 (14) | 102,090 (10) |
| Indonesia | 824,588 | 445,894 | 69,354 (16) | 37,936 (9) |
| Vietnam | 133,918 | 116,391 | 23,598 (20) | 10,233 (9) |
| Sri Lanka | 82,446 | 82,111 | 43,831 (53) | 1,815 (2) |
| Bangladesh | 34,523 | 34,498 | 19,101 (55) | 1,075 (3) |
| All countries | 7,648,792 | 4,191,101 | 545,517 (13) | 210,107 (5) |

*Sources*: On the basis of data provided in Bank of Korea (1995).
*Note*: Figures in parentheses are percentage shares in manufacturing totals.

*International Herald Tribune*, 20 July 1995) and has plans to enter other East European countries; Goldstar has invested in Turkey.

## Taiwan

Since the mid-1980s, Taiwan has also enjoyed current account surpluses and has accumulated enormous foreign exchange reserves. This has led to appreciation of the New Taiwan dollar by nearly 40 per cent against the US$ between 1986 and 1992 alone. The sizeable foreign exchange reserves also led to inflationary tendencies within the economy. Industrial wages increased substantially during this period because of serious labour shortage, and promulgation of labour laws. Between 1990 and 1992 alone, industrial wages had moved up by nearly 35 per cent (Chaponniere, 1992; van Hoesel, 1996). All these factors put together undermined the international competitiveness of Taiwan especially in labour intensive goods which were its traditional exports. To make up for their lost competitiveness, Taiwanese enterprises started moving production to Southeast Asia and China to take advantage of cheap labour reserves. The government enabled this by relaxing controls on capital ouflows in 1987. The ASEAN countries account for over 35 per cent of Taiwan's total approved FDI stock in 1992. The bulk of Taiwan's FDI in ASEAN is concentrated in labour intensive industries, with assembling of electrical and electronic appliances and textiles accounting for nearly 53 per cent of FDI in 1992 in the region (Taiwan, MOEA, 1993). Taiwan has been actively involved in the development of a special economic zone in Subic Bay, the Philippines, and special industrial zones in Vietnam to house Taiwanese companies (van Hoesel, 1996). Taiwanese companies have also made substantial investments in labour intensive activities in mainland China which, however, are not recorded directly because of restrictions and are routed mainly through Hong Kong. According to the host country estimates Taiwanese investments made in mainland China between 1979 and 1992 alone could amount to US$5.5 billion (Zhang and Van Den Bulcke, 1996).

## Hong Kong

Similarly, rising wage costs and scarcity of labour since the mid-1980s forced Hong Kong enterprises to look for locations for labour intensive manufacturing abroad. Mainland China, which had increasingly liberalized its economy since 1979, especially for export-oriented manufacture, proved an attractive location for the relocation of production by Hong Kong firms because of its abundant cheap labour. The geographical, ethnic and cultural proximity and the prospect of the change-over of sovereignty in 1997 further added to the attractiveness of China for relocation. Hence, Hong Kong based enterprises have increasingly concentrated their overseas investments in China since

1985. Hong Kong accounts for about 60 per cent of all contracted FDI inflows in to China since 1979. A considerable part of these investments, however, represents Taiwanese investment in China and possibly also round-tripped Chinese investments which are routed through Hong Kong. Most Hong Kong investment in China is in highly labour intensive assembly operations, producing mainly travel goods, hand bags, toys and footwear in plants located in special economic zones in neighbouring Guangdong and Shenzhen provinces. The bulk of the output of these plants is shipped back to Hong Kong for re-export. The re-export trade constitutes the main dynamic force in the economy of Hong Kong (UN-TCMD, 1993: 38).

## Official promotion of outward FDI flows as a tool for competitiveness in developing countries

In what follows we review the trends in government policies towards outward FDI in a few important developing country sources of FDI in Asia and examine the motivations for their promotion.

### South Korea

The Republic of Korea liberalized its policy governing outward FDI by domestic firms in mid-1980s when its international debt position eased. Since 1989, investments up to $2 million do not require approval. Recognizing the role of outward FDI in promoting international competitiveness of the country's exports, official agencies have promoted it with different measures. The Korean Export-Import (EXIM) Bank gives subsidised loans for overseas investments financing up to 80 per cent (60 per cent since February 1991) of the investment. In the case of medium and small firms, the EXIM Bank financing could even be up to 90 per cent (80 per cent since February 1991). The EXIM Bank and Korean Export Insurance Company also offer investment insurance covering political and non-commercial risks such as war, expropriation or restrictions on remittance to the extent of 90 per cent of the total investment amount. An Overseas Investment Information Centre was established in 1988 primarily to assist smaller enterprises which may not have access to information. The government offers tax incentives such as the reserve for losses incurred by FDI. It also provides for avoidance of double taxation and Korean enterprises can subtract the corporate tax paid abroad from their domestic corporate tax liabilities (UN-TCMD, 1993). In 1992 South Korean companies were allowed to finance their overseas investments completely with foreign loans. However, the pace at which Korean overseas investments have grown since 1990 has raised concerns of accumulation of foreign debt for funding overseas expansion and of hollowing-out of the Korean industrial base with too much relocation of production abroad. This led the Korean Ministry of Finance and Economy to rule in 1995 that Korean

companies must use domestic resources to finance at least 20 per cent of foreign investment in all cases involving investment of $100 million or more
and 10 per cent in smaller projects (*Financial Times*, 10 October 1995).

## *Taiwan*

Taiwanese policy towards outward FDI has also undergone changes over time.
Before 1978, outward FDI was considered unreasonable given the economic
condition of the country. Therefore, 'relatively rigid' financial requirements
were laid down for the qualification of outward FDI and virtually no incentives were offered. The government recognized the importance of outward
investments from the world wide recession and energy crisis in the 1970s,
especially for the supply of natural resources and raw materials for industrial
use and adopted a policy to encourage them since 1979. Outward investments
of an extractive nature for securing supplies of natural resources for domestic
use were granted a five year tax holiday since 20 June 1979. The financial
requirements were relaxed and restrictions on reinvestments were loosened.
A more active promotion of outward FDI began from 1984 when financial
criteria for overseas investment were loosened, the scope of overseas activity
was broadened and the official agencies started financing outward FDI. The
Export-Import Bank of the Republic of China has provided overseas investment financing to the tune of 70 per cent since 1985. In 1986, the Bank
started an investment insurance to cover most non-commercial risks including expropriation, war, riot and restrictions on remittance of capital and profits imposed by host governments. In a major relaxation of controls on capital
outflows in July 1987, the government allowed remittance of up to US$5
million a year out of the country. More relaxed regulations and screening
norms governing outward investments were issued in March 1989. The
Industrial Development and Investment Centre of the Ministry of Economic
Affairs started assisting firms with information, sponsorships of investment
seminars, business group organization seeking investment opportunities and
the establishment of a databank on manufacturers and enterprises interested
in outward investment. In additon, a number of tax incentives for overseas
investments were introduced. Furthermore, the Ministry of Foreign Affairs
provides subsidies to Taiwanese enterprises' investments in Latin American
and Caribbean countries in the form of travel expenses, preferential loans
and insurance. Finally, Taiwan has signed agreements on mutual protection
of investments and avoidance of double taxation with a number of developed
and developing countries (see Taiwan, MOEA, 1993, for more details).

## *China*

Up to 1983 outward investment by Chinese enterprises was modest
and included limited investments made largely in Hong Kong by state

owned foreign trade corporations and financial institutions. Before 1985 the Chinese government hardly paid any attention to the foreign activities of Chinese enterprises. Since then, however, central and local institutions have been set up to regulate overseas activities of Chinese enterprises to gear them to achieve the national objectives – to acquire foreign technology and capital, to promote Chinese exports and to upgrade its national competitiveness in the global markets (Zhang and Van den Bulcke, 1996). The local governments started supporting overseas expansion of their enterprises and foreign trade companies following decentralization of the foreign trade system in 1984. Chinese industrial enterprises have also started making overseas investments since 1985, sometimes in cooperation with foreign trade companies. The investments made by foreign trade companies and industrial enterprises are generally of a different nature. Industrial enterprises operate abroad by establishing new plants, especially in developing countries where their ownership advantages in the form of intermediate production technology and skills could be productively employed. The foreign trade companies have tended to invest in large extractive projects to secure raw material supplies and sometimes have engaged in strategic asset seeking acquisitions of high technology firms in industrialized countries (see Zhang and Van den Bulcke, 1996, for more details).

### *India*

The Indian government has encouraged outward investments by Indian companies as a means of promoting exports of Indian capital goods, technology and consultancy services, especially since 1974 when an Inter-ministerial Committee on Joint Ventures Abroad was created within the Ministry of Commerce to approve proposals. The guidelines for approval were formulated in 1978 which encouraged the joint venture form of operation with local enterprises and required that Indian equity participation be made by way of capitalization of export of indigenous plant, machinery, capital goods and know-how from India. In view of the scarcity of capital resources in the country, cash remittances of capital to overseas ventures were discouraged but could be allowed in exceptional cases. This policy governed Indian outward FDI till the end of the 1980s. As part of a comprehensive reform of India's industrial policies initiated in July 1991 the government also removed some of the restrictions on Indian outward FDI (see Kumar, 1994c, for more details). The modified Guidelines for Indian Direct Investment in Joint Ventures and Wholly Owned Subsidiaries Abroad issued in October 1992 provide for an automatic approval for proposals where the total value of Indian investment does not exceed US$2 million, of which up to US$ 500,000 could be in cash and the rest by capitalization of Indian exports of plant, machinery, equipment, know-how or other goods and services. The approval for other proposals including external borrowing, use of export receipts blocked abroad etc. will be made within ninety days after due consideration with respect to the track record of the Indian party in terms of external orientation and financial viability of proposed investments etc. The Export-Import Bank of India provides financing for outward FDI by Indian enterprises under its Overseas Investment Finance Programme. However, investments of more than $15 million should preferably be financed by raising resources abroad with share issues by the Indian companies. As a result of this liberal outward investment policy, Indian companies are reported to have invested $200 million abroad during 1995–6 (*Economic Times*, 24 August 1996).

## Developing countries as sources of FDI inflows for poorer countries: future prospects

From the above account it is clear that developing countries have emerged as significant sources of FDI outflows especially for other developing countries. Although FDI flows originating in the developing countries currently account for a 15 per cent share of global inflows, their share in inflows to developing countries is much more substantial. The detailed patterns of outward FDI of all sources of FDI among the developing countries are not readily available. However, according to our computations based on the available data, investments originating in developing countries account for 35 per cent of all FDI inflows received by the developing countries. Even if rather large ethnic investments from Hong Kong to China are ignored, developing countries would account for at least a quarter of FDI inflows received by all developing countries (excluding China). Developing countries constitute a much more significant source of FDI for them than would appear from the global figures despite the recent trend of increasing concentration of developing country outflows in industrialised countries.

The fact that developing countries are important sources of FDI for co-developing countries is also clear from Table 7.6 which provides the shares of inter-developing countries' FDI in the total inward FDI stock in a number of Asian and Latin American countries. FDI inflows originating in developing countries evidently hold an important place in a number of developing countries. For instance, FDI flows from developing countries accounted for 65 per cent of inward FDI stock in China in 1990, nearly 50 per cent in Sri Lanka, 41 per cent in Malaysia, 37 per cent in Paraguay, and was approaching 30 per cent in Indonesia, Chile and Taiwan. The significance of developing countries as sources has grown in more recent years. For instance, East Asian developing countries were responsible for 50.6 per cent of all FDI approvals in 1990–1 in Indonesia and 44.6 and 25.6 per cent of FDI approvals in Thailand in 1990 and 1991 respectively (Wells, 1993). Tho (1993: table 9.8) similarly shows that East Asian countries account for 47 per cent of cumulative inflows in Malaysia between 1988 and 1990 and 16 per cent in Thailand in 1988–9.

With increasing relocation of production to cheaper labour locations for strengthening price competitiveness of their goods by enterprises based in East Asian countries as shown in the fourth section above (pp. 184–7), these countries are also emerging as significant sources of export-oriented production. It has been shown that a substantial proportion of investments from these countries in ASEAN and South Asian countries has gone in to export oriented industries. Wells (1993) reported that 46 per cent of all FDI projects originating in East Asian countries approved in Indonesia between 1990 and 1991 were export-oriented compared to 30 per cent in the case of FDI projects originating in industrialized countries other than Japan. The importance of developing countries as sources of FDI is likely to increase as more enterprises from developing countries employ outward investments as a strategic tool for strengthening their presence abroad in the coming years with the increasing protectionism and growing emphasis on internationalization of operations. For instance, the top five Korean chaebols alone had planned to invest $70 billion overseas in the next decade (*Economist*, 14 September 1996).

A considerable volume of literature in the early 1980s analysed the relative characteristics of FDI from developing countries and brought out a number of positive features from the perspective of the host country. It was stressed that the main ownership advantages of developing country enterprises consisted of adaptations made to process technology and other operational parameters to better suit the developing country situation. This resulted in a smaller scale of operation of developing country enterprises which were more appropriate

*Table 7.6*  Share of FDI originating in developing countries in FDI stock in host countries, 1980–90 (percentage)

| Latin American countries | 1980 | 1990 | Asian countries | 1980 | 1990 |
|---|---|---|---|---|---|
| Argentina | 4.5 | 5.9[a] | Hong Kong | – | 16.9[a] |
| Bolivia | 15.1 | 17.3 | Republic of Korea | 8.2 | 5.8[a] |
| Brazil | 10.0 | 7.3 | Singapore | 11.5 | 5.4[a] |
| Colombia | 17.0 | 9.4 | Taiwan | 36.8 | 27.7[c] |
| Ecuador | 27.8 | 25.5 | Indonesia | 22.9 | 27.9[c] |
| Chile | 12.8 | 29.1 | Malaysia | 41.4[g] | 40.8 |
| Mexico | 0.5 | 5.6 | Philippines | 8.0 | 9.4[h] |
| Paraguay | – | 37.0[c] | Thailand | 20.3 | 22.8 |
| Peru | 15.5 | 22.9 | China | 41.5[i] | 65.0[h] |
| Uruguay | 7.1[d] | 26.2[a] | Vietnam | – | 16.1[a] |
| Venezuela | 17.4 | 12.5 | Bangladesh | 2.1 | 13.4[h] |
| El Salvador | 25.1 | 24.6 | Pakistan | 19.5 | 27.2[c] |
| Guatemala | 52.3 | – | Sri Lanka | 45.2 | 49.1 |

*Sources*: Compiled from UNCTC, 1992, *World Investment Directory*, Volume 1: Asia and the Pacific; and UNCTAD, 1994, *World Investment Directory, Volume 4: Latin America and the Caribbean.*
*Notes*: a 1989; c 1988; d 1978; e 1986; f 1977; g 1981; h 1987; i 1982.

for the smaller markets of host countries and led to better utilization of capacity. Adaptations made to technologies made them not only more appropriate for the host economy compared to those by foreign enterprises originating in industrialized countries in terms of factor proportions but also to the quality of local infrastructure and implied greater use of local raw materials and skills, and lower consumption of foreign exchange per unit of output (see among others, Lall *et al.*, 1983; Wells, 1983; Agarwal, 1985; see Kumar, 1986 for a review). Because these ownership advantages could provide an edge *vis-à-vis* local enterprises only in countries that were situated below their home country in terms of technological and economic development, the bulk of FDI in the early period was concentrated in such developing countries. However, as developing country enterprises, especially those from East Asian NIEs, have matured some of the early positive characteristics of their investments such as intermediate nature of technology may have become less applicable. Wells (1993), for instance, reported that the differences between developing country and industrialized country based enterprises in Indonesia tended to narrow over time as the former moved increasingly into export-oriented manufacturing. Nonetheless, low income countries and least developed countries which have been marginalized in the global distribution of FDI inflows may find it easier to attract FDI from developing country enterprises than from industrialized countries. For instance, compared to a just about 1 per cent share in global FDI inflows, African countries have received 3 per cent of South Korean outflows and nearly 10 per cent of Indian FDI flows. Second, export-oriented FDI flows which presently constitute a bulk of Asian NIEs' overseas FDI in ASEAN countries are known to be of a footloose nature. They move away with the erosion of the current host country's locational advantages such as cheap labour. It is expected that export-oriented FDI originating in the East Asian NIEs will diffuse more evenly across developing countries in the future with the saturation of ASEAN countries as bases for labour intensive processing.

Two factors which constrain potential inter-developing country FDI flows especially to the relatively poorer or least developed countries are the lack of information and financing. Information on the potential investment opportunities arising in the least developed countries is hard to come by and developing country enterprises, barring a few, lack global information systems of the type conventional MNEs are known to be equipped with. Finally, a few emerging MNEs can match the financing ability of the conventional MNEs. Hence, institutional intermediation at the national, regional or international level may be instrumental in directing these flows to the least developed countries.

The attempts at regional economic cooperation among developing countries would also facilitate inter-developing country FDI flows and technology transfers. Participation in regional economic integration, besides promoting intra-regional investments, could help developing countries attract FDI originating in developing countries outside the region by increasing the effective market size accessible. For instance, Indian companies are reportedly working on twenty-six joint ventures in Mauritius and other African countries following the Common Market of Eastern and Southern African States (COMESA) covering a clearing and financing mechanism and 50 per cent import duty concession on intra-regional trade (*The Financial Express,* 15 August 1996). Regional economic integration could also help to create some institutional infrastructure at the regional level to fill the informational and financing gaps. Finally, agencies promoting FDI inflows in poorer developing countries would be well advised to target FDI originating in developing countries. Wells (1993) observes that some Latin American countries, e.g. Costa Rica and Colombia besides the Southeast Asian countries have recognized the potential of attracting export-oriented FDI from East Asia and have begun to tap it successfully.

## Concluding remarks

The foregoing discussion has highlighted an aspect of growing internationalization of the world economy in the recent period – the increasing resort by developing country enterprises to direct investments abroad as a strategic tool for strengthening their competitiveness. The erosion of competitiveness caused by currency appreciations and rising wages has been addressed by relocation of production to countries with lower wages. The threat of losing markets in industrialized countries because of rising protectionism in the wake of formation of regional trading blocks has been responded to by making trade supporting and strategic asset seeking investments in major markets. The emerging MNEs have been assisted in their overseas expansion by the policy framework and infrastructural support of their home governments, given their positive role in strengthening their international competitiveness.

The developing country enterprises have emerged as important sources of FDI inflows especially for developing countries for whom they provide over a quarter of all inflows. Their importance as sources of FDI inflows is strikingly high for the ASEAN and some Latin American countries. The emergence of developing country enterprises as important outward investors is a significant development of the past fifteen years. It widens the options of developing countries looking for FDI inflows and technology at least in standardized and matured industries. Emerging MNEs have often provided to their lesser developed host countries access to intermediate and generally more appropriate and cheaper technologies and skills. This intermediate range of technologies could be valuable in the development of wage goods sectors in dual economies. Developing countries are becoming sources for not only domestic market oriented FDI but also for export-oriented ventures. The least developing countries may find it easier to attract FDI originating in developing countries than in industrialized countries. It is expected that the inter-country distribution of FDI emanating from developing countries would be more even in the future with saturation of ASEAN countries as bases for labour intensive processing. Yet there appears to be scope for institutional intermediation at the national, regional or international level for directing these flows to the poorer countries given the constraints of information and financing. On the part of receiving countries, a specific targeting of developing country FDI may be desirable. Some countries have already begun to successfully target them. The efforts at regional economic cooperation among developing countries could also facilitate inter-developing country FDI flows and technology transfers.

# Part IV

# DEVELOPING COUNTRY PROSPECTS AND POLICY IMPLICATIONS

Finally, this part draws out the threads from the previous three parts on the prospects and implications for developing countries of the increasing globalization with expansion of FDI and other cross border activity. It examines prospects for increased FDI inflows to developing countries. It discusses the role of factors that shape the distribution of FDI among the developing countries and hence implications for policies of the least developed countries. Then it outlines certain aspects of strategic interventions by developing country governments to maximize the gains from globalization and a few important issues of international action in the area of international business.

# 8

# FOREIGN DIRECT INVESTMENTS AND TECHNOLOGY TRANSFERS IN AN ERA OF GLOBALIZATION AND PROSPECTS FOR DEVELOPING COUNTRIES

## A policy postscript

*Nagesh Kumar*

## Introduction

As a part of the globalization of economic activity, the cross-border transactions of technology and direct investments have expanded greatly over the past two decades, especially since the mid-1980s. Several global economic events of a reinforcing nature have contributed to this expansion. These trends include: worldwide liberalization of national economies to trade and investments; privatization; regional economic integration in many geographic regions – especially in the EU and North America and the corporate restructuring sparked off by them; the emergence of new generic or core technologies; reforms in the Central and Eastern European countries; appreciation of the yen since the Plaza Accord; and rapid technological learning and industrialization in the East Asian countries, among others. The annual magnitude of FDI flows has risen from about $25 billion in the early 1970s to nearly $350 billions in 1996, and that of annual flows of technology transfers in terms of international technological payments from $7–8 billion to over $60 billion over the same period. This expansion in cross-border activities of enterprises has been accompanied by much restructuring of the global pattern of production, and hence, technology, trade and investment – where some countries have gained at the cost of others. The extent of the restructuring or 'reconfiguration' as Dunning puts it, over a short period of two decades is probably more profound than the changes taking place over the past two centuries.

The contributions in this volume have analysed the trends in the international technology transfers and direct investments through their metamorphosis over the 1975–95 period especially to examine their implications and prospects for developing countries. Since the individual chapters have generally summarized their findings at the end, here we shall confine ourselves to recapitulation of only those having significant implications for the industrialization policies of developing host countries.

## Emerging patterns of technology generation and transfer

The global pattern of technology generation shapes the patterns of international technology transfers and direct investment flows. The technology generation activity is found to be highly concentrated in a handful of advanced industrialized countries with the US, Germany and Japan accounting for over four fifths of patents granted by the US Patents Office, although there seems to be a slightly more even distribution of generation between them over time. Barring a couple of newly industrializing countries such as Taiwan and South Korea, there does not seem to be much prospect of a developing country emerging as a serious contender in technology generation in the near future. In fact the technological effort in developing countries as a group in terms of global R&D expenditures has declined and the technology gap between them and industrialized countries has widened over time rather than narrowed. The hold of larger corporations on the generation of new technologies seems to have intensified with the growing scale of R&D activity and recent trend of corporate consolidation and restructuring through mergers and acquisitions, strategic alliances and a variety of industry–university–government complexes, especially in the new core technologies. A strong interaction between corporate size, R&D expenditure, patent ownership and multinational orientation is apparent. These interactions are expected to grow stronger with increasing globalization of the world economy and new international norms of protection of intellectual property and may have adverse implications for the terms of technology transfer especially for the poorer countries.

In terms of the mode of technology transfer, the trends suggest a reversal of the growing popularity of arm's length licensing in the period between mid-1970s and mid-1980s. Since the mid-1980s, market or arm's length transactions have steadily lost their importance and intra-firm transfers – those under the package of FDI – have regained their prominence as the mode of technology transfers. Over 80 per cent of transfers by US corporations (in value terms) and nearly 95 per cent of transfers by German corporations have been made on an internal basis in 1995 compared to 69 and 92 per cent in 1985 respectively. The proportions for British technology transfers have gone up from 52 per cent to over 76 per cent between 1985 and 1992. Japanese corporations transfer a much more significant proportion of technologies (in terms of numbers) through markets but the proportion of intra-firm transfers has risen even for them. The increasing domination of the intra-firm mode of technology transfer results from, to some extent, an increasingly receptive attitude among the technology importing nations to the FDI as reflected in the liberalization of investment codes world-wide. Partly, it may reflect the growing reluctance of larger corporations that increasingly control much of technology generation to part with their technology unaccompanied by ownership and control. This is because growing internationalization of markets over the past years has emphasized the role of technology as a key element of international competitiveness. In the terminology of the eclectic theory of overseas operations of firms, the internalization incentives in the market transactions of technology have considerably increased in the age of globalization of markets because of increased economies of scope. The recognition of the role of technology as an instrument of competitiveness has even prompted the governments of industrialized countries to support and protect the technological effort of national enterprises especially in new core technologies in a manner that has been termed 'technonationalism' and 'technoprotectionism' (see Kumar and Siddharthan, 1997).

Despite the bulk of technology transfers taking place on an intra-firm basis, a close look at the rates of expansion of disembodied technology transfers and FDI inflows gives the impression that the recent expansion of FDI inflows has not been accompanied by technology flows in the same proportion or that the technological content of FDI flows has declined over time. This has been explained in terms of the fact that much of the recent expansion in FDI has been accounted for by corporate restructuring in the industrialized world and expansion of FDI in services rather than greenfield investments in manufacturing. Although FDI in services and mergers and acquisitions may involve substantial transfer of knowledge across borders, it may be smaller than that involved in the former case on a per unit of invested capital basis.

The triad nations – USA, European countries and Japan – also absorb among themselves the bulk of technologies transferred across countries. The technology transfers from Europe and Japan reveal a strong regional focus. Although the share of developing countries in global technology transfers has gone up in recent years, the long-term trend has been one of decline. Finally, the bulk of technology transfers to developing countries are actually received by a handful of relatively more developed and technologically more dynamic countries and this concentration is increasing. From this, the local absorptive capacity would appear to be one of the determinants of a country's ability to receive technologies from abroad.

## Emerging patterns of FDI inflows and their explanations

FDI flows not only serve as a channel for nearly four fifths of global transfers of disembodied technology, they have also emerged as the most important source of long-term external resource flows to developing countries since 1992 because of stagnation of official development assistance. Sometimes FDI inflows also assist their host countries with expansion of exports by relocating production geared to their home markets or to third countries. FDI is also seen as a channel of diffusion of new management and organizational techniques in the host countries and as improving the efficiency of local firms with competitive effects and spillovers of knowledge. Hence, there is increasingly intense competition among countries to woo FDI inflows through liberalization of policy regimes, investment and tax incentives and other attractions. However, some countries have been able to benefit more from the recent expansion of global FDI inflows than others. The emerging global patterns of distribution of FDI were analysed to get insights into the factors explaining them. It was followed up by a more detailed analysis of patterns emerging for the major source countries of FDI.

Several changes are notable in the geography of FDI over the past two decades. These include the emergence of China as one of the largest recipients of FDI world-wide. Actually the surge of inward FDI to China explains in large part the increasing share of developing countries in global distribution of FDI in the 1990s. Excluding China, there has been a sharp reduction in the concentration of FDI among developing countries. Within the North and South, some regions have gained at the expense of others. For instance, the Southeast and East Asian newly industrialized economies have become more attractive hosts relative to most countries in Latin America. Western Europe, especially France and Spain, have gained as FDI recipients at the expense of Canada and the USA. Central and Eastern Europe, especially the Czech Republic, Hungary and Poland have begun to emerge as quite important recipients, while Africa continues to be of marginal interest to foreign investors. The traditional resource based recipients of FDI, – Canada and Australia – have lost ground to the faster growing industrialized countries.

To a considerable extent, changes in the geography of FDI reflect the changing pattern of domestic investments in the host countries. However, the growth of inward FDI has been considerably faster than that of domestic investment in France, in Canada and in parts of South and East Asia (especially China) and Central Europe and has been least in West Asia and most of Africa.

Dunning would explain the changes in the geography of FDI specific to its foreignness in terms of the impact of global political and economic events on the configuration of ownership advantages of firms, locational advantages of countries and internalization advantages of firms (or OLI advantages) facing foreign investors. The regional and global competition between firms is increasingly replacing national competition in the 1990s as a result of worldwide liberalization of economies; deregulation of markets and privatization of enterprises; regional economic integration in many regions; technological advances which have lowered the costs of communication; acquisition of information and of transportation; and corporate restructuring with an emphasis on competitiveness. In this scenario, the conventional ownership advantages in the form of privileged access to resources and markets and other country specific knowledge turn out to be less important compared to those associated with multinationality *per se*; and the ability of firms to seek out and exploit complementary assets by cooperative arrangements with their suppliers, competitors and customers across the world. These networking strategies enable the participating firms to take maximum advantage of the liberalization of markets, regional integration and new technological advances by complementing and upgrading their core competencies.

Similarly, conventional locational advantages of host countries based on natural resources (with increasing dematerialization of production) and cheap labour endowments (with increasing employment of flexible auto-mation systems in production), protection, taxes and incentives are becom-ing relatively less valuable compared to the availability of knowledge creating activities which help foreign investing firms exploit and augment their competitive advantages. Finally, internalization incentives in the cur-rent era arise more from the need to spread political and business environ-mental risks across the world, from a holistic integration of disparate functions and strategies and from economies of scope, compared to conven-tional market failures involved in transfer of intangible assets. These changes have guided the pattern of the recent spate of strategic asset seeking invest-ments in the form of cross-border M&As which account for a substantial proportion of recent expansion of FDI flows – especially those between the triad countries.

Another source of changes in the geography of FDI is the changing compo-sition of home countries, the nature of the economic activities undertaken, and the structure and strategies of the participating firms. The most signific-ant growth in outward FDI over the past two decades has been recorded by MNEs from Japan and from Asian NIEs benefiting South, Southeast and East Asian countries; on the other hand, the US MNEs achieved only a mod-est growth in their foreign activities which adversely affected the share of Latin American countries in FDI. The changes in the sectoral composition of economic activity have had a more ambiguous effect on the geography of FDI. The growth of the technology and knowledge intensive industrial sec-tors and high value added services have encouraged more FDI in the most advanced economies; opening of many infrastructural services has also led to FDI in developing countries.

## Prospects for developing countries

What prospects do the expanding magnitudes of technology transfers and FDI inflows and their changing geography create for developing countries? The recent expansion of international technology transfers and FDI inflows has been accompanied by their greater concentration in the industrialized countries. The share in FDI inflows between mid-1970s and mid-1990s of developing countries as a group, excluding China which has emerged as a major host country of FDI in the 1990s, has declined. The share of developing countries in global transfers of disembodied technology also shows a declining trend although it appears to have risen over the past few years. The bulk of FDI inflows directed to China have been largely driven from ethnic and cultural reasons from within the region and hence are of a different nature from conventional inflows. The 100 smallest recipient countries of FDI inflows (including all the least developed or the poorest countries) have collectively shared only 1 per cent of these flows – about $3 billion (UNCTAD, 1996: 4). Therefore, the recent booms in cross-border investment activity are driven by industrialized countries (UNCTAD, 1997: 10–1). The analysis of emerging trends in the internationalization activity in major source countries also does not suggest any imminent boom in the near future.

## US MNEs

Lipsey, in Chapter 4 of this volume, found the overseas production ratio of US MNEs to be stagnating over the past two decades. Developing countries' share of internationalized US production fell between 1977 and 1991 and then recovered somewhat, reflecting the shift in US FDI from resource-oriented to market-oriented investment. The internationalized manufacturing production of US MNEs remains concentrated in developed countries. The share of American firms in total production and manufacturing output in developing countries as a whole and in most individual countries except in the Asia-Pacific region has also declined. In view of this, Lipsey feels that internationalized production of US MNEs is unlikely to be a major source of developing country economic growth despite the current globalization rhetoric. He also did not find a statistically significant causal relationship between US owned production in the countries and total output and growth in per capita income in the subsequent period.

## European MNEs

Technology transfers and FDI outflows from European countries have increasingly been concentrated within Europe and its peripheries over the past decade or so with regional economic integration. It is evident from Table 8.1 that the FDI outflows of EU member states became increasingly concentrated within the region after the Single European Market was put in motion with the White Paper of 1987, bringing the share of intra-regional flows in outflows to 71 per cent by 1990. The analysis of Chapter 2 has also shown that 52 per cent of transfers of disembodied technology by European countries was also absorbed by the EU and EFTA partners. Western Europe and the US, which also has been a major destination of the EU's outward investments, together absorb 85 per cent of the EU's FDI outflows. The remaining FDI outflows of EU member states are increasingly concentrated in the Central and Eastern European countries, and in the Mediterranean countries which enjoy privileges under the so called 'Europe Agreements'. Agarwal in his chapter has observed the strongest growth of German FDI outflows to be in the Central and Eastern European countries. Relatively untapped markets, geographical proximity, low labour costs, privatization, preferential trading arrangements and cultural relations have facilitated the involvement of German MNEs in this region. Agarwal expects the German and other European FDI to rise more than proportionately in Central and Eastern Europe in future because of the relatively small base of FDI in these regions. The emergence of Central and Eastern European and Mediterranean countries as hosts for European MNEs has indeed increased locational competition for developing countries in other continents, limiting their prospects for attracting FDI flows originating in Europe in the future.

*Table 8.1* Distribution of FDI outflows of European Union members, 1985–93 (ECU million)

| Destination | 1985 | 1987 | 1990 | 1993 |
|---|---|---|---|---|
| Intra-EU Outflows | 5,949 | 12,344 | 39,295 | 30,844 |
| Percentage share in total | 28 | 29 | 66 | 59 |
| Extra-EU outflows | 15,105 | 30,670 | 20,527 | 21,854 |
| of which to EFTA countries | 722 | 1,789 | 3,226 | 4,028 |
| Total intra-EU and EFTA | 6,671 | 14,133 | 42,521 | 34,872 |
| Percentage share Western Europe | 32 | 33 | 71 | 66 |
| USA | 10,061 | 23,885 | 7,155 | 10,167 |
| Total Europe and USA | 16,732 | 38,018 | 49,676 | 45,039 |
| Percentage share Europe and USA | 79 | 88 | 83 | 85 |
| Total outflows of EU countries | 21,054 | 43,014 | 59,822 | 52,698 |

*Source*: Based on EUROSTAT (1995), statistical tables, pp. 158–9.

## Japanese MNEs

Japan has fast emerged as one of the most important sources of technology and FDI flows over the past decade. Japanese FDI outflows are rising again after a brief period of decline in the early years of the 1990s. The appreciation of the Japanese yen has provided the major push factor, besides the accumulation of managerial and technological capabilities by Japanese firms. Japanese MNEs have a greater proportion of their cross-border technology transfers, as well as investments, in developing countries, than the US and European

MNEs. Rising protectionist tendencies in Europe and North America directed against Japanese exports, in the form of voluntary export restraints, screw-
driver regulations, etc. and the emerging threat of discrimination against
extra-regional supplies after the completion of regional economic integration
pushed up the share of these regions in Japanese FDI outflows in the second
half of the 1980s. However, Asian countries have received a substantial proportion of Japanese manufacturing FDI. In this region, robust economic
growth, low wage rates and liberalization of trade and FDI policies acted as
pull factors. Although the recent depreciation of the yen has slowed down
the pace of Japanese FDI, Urata expects the further internationalization of
Japanese corporations to continue with a predicted 85 per cent increase in
overseas production ratio in the next fifteen years. MNEs will continue to
be pushed to move production abroad by labour scarcity in Japan, by the
need to achieve efficiency, by internationalization of production and by the
increasingly receptive attitude of governments towards FDI inflows. Some
of this internationalized production, especially the cheap labour seeking
type, can be expected to be hosted by developing countries. Urata also observes
a shift in the focus of Japanese FDI especially in the electrical machinery industry from East Asian NIEs during the period 1986–8 to ASEAN countries in
1989–92 and to China in 1993, as predicted by the flying geese theory. He
finds signs of other countries such as India and Vietnam starting to attract Japanese FDI in electrical machinery. So the Asian countries with low wages but
quality manpower and with reasonably good infrastructure and other facilities
(besides those with sizeable and growing domestic markets) could expect to
host relocated production by Japanese MNEs.

### Emerging sources

Although Asian NIEs and other emerging developing countries are relatively
new entrants as significant outward investors, they have already emerged as
important sources of FDI for developing countries. They account for between
25 and 33 per cent of inflows received by developing countries in the mid-
1990s depending upon whether Hong Kong's outflows to China are excluded
or included. Their importance as sources of FDI inflows is strikingly high for
the ASEAN and some Latin American countries. Developing countries are
becoming sources for not only domestic market oriented FDI but also for
export-oriented ventures. The currency crisis of 1997 has affected the economies of Southeast and East Asian countries and will temporarily slow down the
investment plans of corporations based in these countries. However, one
would expect that their importance as sources of FDI will grow further,
after overcoming the current crisis, given the tendency of enterprises based
in these countries to internationalize production in order to strengthen
their international competitiveness. This is because the internationalization
of production by enterprises based in these countries has been pushed by

their substantial technological learning which has revenue productivity in other countries, by the exhaustion of their MFA and GSP privileges, by growing labour scarcity, etc. rather than their financial strength. Most of these reasons continue to remain valid. Furthermore, regional economic integration among developing countries, when it takes off in an effective manner, will create many opportunities for intra-regional FDI flows in an attempt to rationalize and restructure production across the region. A lot of potential for inter-developing-country FDI and technology transfers, such as in intermediate technologies, and for technologies adapted to specific developing country markets, or environmental situations, remains unexploited for want of information and financing. With some institutional intermediation in those spheres, it may be possible to promote these. The emergence of the new sources of technology and FDI widens the options of developing countries looking for FDI inflows and technology at least in standardized and matured industries. The least developing countries may find it easier to attract FDI originating in the emerging countries than from conventional sources in the industrialized countries. These investments may not only be more easily attracted, but may have more positive developmental effects, as some research conducted in the 1980s and cited in the previous chapter had shown, because of their more appropriate scale and technology, among other differences.

To sum up this section, there appears to be little prospect of a major boom in FDI-led expansion in output and growth in developing countries despite the recent dramatic expansion of global inflows of FDI. The two poles of the triad of FDI sources – US and EU – mutually absorb the bulk of their FDI outflows and the trends suggest no change in this tendency in the near future. Therefore, Japan, Asian NIEs and other emerging sources of FDI hold the key to expansion of FDI inflows to developing countries. The turbulence caused by the currency crisis of mid-1997 has affected some of the emerging source countries from Hong Kong and South Korea to Malaysia, Indonesia and Thailand rather badly. This may adversely affect the outward investments of enterprises based in these countries. Weakening of the yen and other countries' currencies may also adversely affect the plans for relocation of production away from home bases. However, this appears to be a temporary disruption and as these economies pull themselves out of the crisis and stabilize their financial sectors with reforms, the trend of outward investments is expected to be revived.

## Factors shaping the international distribution of technology and FDI: implications for policies

The analysis of the previous chapters does provide some pointers on the future distribution of FDI inflows among developing countries. Below we discuss a few implications for the policy of developing countries especially those having difficulty in attracting FDI inflows and technology arising from the above. The findings of a more detailed quantitative analysis on some of these issues are reported in a companion volume (Kumar, forthcoming).

The major determinant of FDI appeared to be the presence of local markets resulting in its increasing concentration in industrialised countries, as observed by Dunning overall as well as by the country case studies. The US study found even the production for export concentrated in the countries with large domestic markets. Per capita income appeared to be a major influence among the factors that affect the location of FDI, which explains the declining competitiveness of least developed countries, such as those in Africa, in attracting FDI and technology through markets, as emphasized repeatedly in Chapters 2, 3, 4 and 6. This situation prevails despite the quite favourable policy framework for FDI inflows in many of these countries. This finding corroborates the results of a number of quantitative studies which have reached converging findings on the factors favouring FDI inflows despite the divergent samples, measurements of variables and methodologies used. These factors include per capita income levels, market size, growth rates, extent of industrialization and urbanization, quality of infrastructure and macro-economic stability, among others. The policy factors such as degree of openness of the economy, incentives, tax rates, etc. have generally not proved significant determinants of FDI inflows. Contractor (1990) in his empirical study of forty-six countries did not find liberalization to be an important factor in influencing the pattern of FDI flows. The foreign investors' response was found to be strongly influenced by the size and growth of the host economy rather than by changes in the government's FDI policies. Wheeler and Mody (1992) in their study covering forty-two countries for the period 1982 and 1988 emphasized the importance of the quality of infrastructure, level of industrialization and market size in attracting US FDI. The open market policies or incentives, such as tax breaks, were found to be of limited value in determining the investment decisions of US MNEs. Dunning finds the importance of traditional sources of locational advantages such as cheaper labour and access to natural resources and raw materials to be diminishing with increasing globalization of the world economy. This would suggest that the low income and least developed countries may be even more marginalized in future in terms of distribution of FDI inflows because the locational advantages of this group of countries are often based on cheap labour and natural resource endowments; so the concentration of FDI inflows in to high and middle income countries will intensify. China is the only low income country that has managed to attract large magnitudes of FDI in recent years. But it has been noted that FDI in China is of a very special type dictated by cultural factors more than economic and has been largely sourced from Hong Kong, Taiwan, Macau and Singapore rather than traditional sources of FDI. Other low income and least developed countries do not enjoy the privileged access to ethnic entrepreneurial communities abroad with substantial investible resources.

The above discussion raises a question mark regarding the ability of the least developed countries to attract substantial magnitudes of FDI inflows, especially from the conventional Western sources, despite their recent expansion. This suggests that the expansion of FDI inflows, while a welcome development by itself, can hardly compensate for the shrinking levels of official development finance, both in nominal as well as in real terms over the recent years, which are generally expected to be directed to the poorer countries.

Hence, the least developed country governments would do better by focusing on improving infrastructure, human resources, developing local entrepreneurship, creating a stable macroeconomic framework and creating conditions conducive for productive investments to get the process of development started rather than trying to attract FDI with incentives and policy liberalizations. The restoration of concessional development finance should help by complementing the meagre domestic investible resources. As we observed in Chapter 3, a considerable part of the variation in FDI across countries could be explained in terms of the geography of domestic investment. Once the pace of industrialization picks up, FDI will probably flow in by itself. Similar sentiments have also been expressed elsewhere. For instance, Loree and Guisinger (1995) found no evidence to suggest that 'a dollar spent on incentives has a higher return, in the form of investment attracted, than a dollar spent on infrastructure'. Agarwal *et al.* (1991: 128) finding the irrelevance of specific incentives and attractions offered to foreign investors concluded that what is good policy for domestic investors – for instance, a stable and favourable general framework for investment – is also good for foreign investors.

The other implications for policy arising from this analysis of some of the factors affecting distribution of FDI and technology transfers are described briefly in the next four sections.

*Building local technological capabilities needs to be accorded due emphasis in developing countries' development policy*

Local technological and absorptive capacity determines the ability of a country to import technology and to use the imported technology effectively. Technological capability's importance as a locational advantage of potential host countries has increased further with globalization. The R&D spending in developing countries, except for Asian NIEs, has declined over the past decade from already low levels as observed in Chapter 2. This neglect of technological effort will adversely affect their ability to mobilize new technologies – local or imported – in their process of industrialization. In this connection it is important also to keep in mind the cumulative nature of technological learning and hence increasing returns to scale in technological activity which are now widely recognized in the new growth theory literature (see Kumar and Siddharthan, 1997, for reviews).

## *Protectionist policies still promote market seeking FDI inflows*

The theory of international business perceives the protection accorded to local production from imports creating locational advantage and forces the exporting firms to undertake market defensive local production. A number of studies have found a positive relationship between the extent of protection and the extent of FDI inflows (see Dunning, 1993; and Caves, 1996, for reviews of evidence). In more recent times, EU countries have improved their competitiveness as hosts of FDI inflows over the years with increasing protectionism e.g. by imposing voluntary export restraints and screw-driver regulations on Japanese and Korean exports. Outside MNEs (e.g. Japanese and now Korean) were prompted to undertake market defensive FDI in response to these regulations. Furthermore, the threat of discrimination against extra-regional supplies following the formation of the Single European Market has prompted American, Japanese and other foreign suppliers to undertake FDI within the EU countries and claim insider status (see Kumar, 1994b, for illustrations). As a result the EU countries increased their share in global FDI inflows from about 26–27 per cent in the 1980–7 period to nearly 50 per cent in the early years of 1990 as the 1992 deadline approached. This suggests that given the market size, protectionism continues to be a quite significant determinant of FDI inflows, especially the market defensive ones. The excessive liberalization of trade regimes by developing countries as a part of structural adjustment programmes may take away some of their attractiveness especially for market seeking investments, as MNEs may decide to export rather than produce locally.

## *Geographical and cultural proximity are important factors in FDI location*

The importance of cultural and geographical proximity in determining the pattern of location of FDI flows has been emphasized in the different chapters, especially in the German case study. These factors have also influenced the pattern of intra-regional FDI flows in Asia as observed in Chapter 7. With the new organizational techniques, such as just-in-time methods of component inventories being increasingly implemented by MNEs world-wide, the importance of geographical proximity will rise further – especially for offshore processing or OEM component manufacture. This again highlights the importance of tapping the potential of intra-regional FDI and technology transfers.

## *Regional economic cooperation among developing countries may help them exploit the potential of intra-regional FDI and gain from efficiency seeking restructuring of industry*

Since the size of host country market turns out to be an important determinant of FDI inflows, the developing countries (especially the smaller ones) could improve their chances of attracting FDI inflows through participation in regional economic integration schemes which has the effect of enlarging the size of the accessible market. Furthermore, regional economic integration among developing countries may also help them to exploit the potential of intra-regional investments which may be substantial and may help in efficiency seeking restructuring of industry in the participating countries.

## Strategic interventions to maximize advantages from globalization

The contributions in the book also provide certain inferences for policy interventions – opportunities for taking greater advantage from globalization by developing countries, especially those having some capability. These include support to national champions, a selective approach to FDI, competition policies and policies for diffusion of knowledge brought in by MNEs.

### *Strong national champions enable countries to reap the benefits of globalization*

MNE investment has favoured locations that provide complementary location bound created assets especially for knowledge intensive value adding activities. Countries that are able to create what Dunning (in Chapter 3) calls 'a strong nucleus of flagship national firms' and what others call 'national champions' (Kumar and Siddharthan, 1997, for instance), will be better placed to tap the MNEs' value adding resources than others, especially in the internationally oriented sectors. The national champions can enter into mutually beneficial alliances with foreign multinationals and could also give visibility to national capabilities in those industries.

Governments in industrialized countries the world over pursue policies to nurture, support and protect national champions in strategic industries by a combination of policies as a part of strategic trade policies. These policies cover R&D subsidies, preferences in public procurements and using bilateral aid and diplomatic resources to support them in overseas business (see Scherer and Belous, 1994, for more details). The national champions are protected from the threat of take-overs by foreign companies in a number of countries.

Governments in developing countries may take some lessons from these policies. These strategic interventions, depending upon the existing conditions, could take the form of consciously promoting national enterprises, assisting them in the process of building technological capability by fostering linkages with R&D institutions and by giving protection from the threat of take-over, especially in the strategic sectors. More specific formulation of the policy will have to be made in the specific context of a country.

## *Selective FDI policies may provide conditions for an autonomous path of expansion*

The experience of East Asian countries has highlighted the fact that selective approach to FDI may provide a necessary condition for charting an autonomous path of internationalization. Evidently, the East Asian newly industrializing economies have consciously sought technology – as far as possible unaccompanied by ownership – with a selective policy. As documented very well in the literature (see, for instance, Kim, 1997; Kumar and Siddharthan, 1997), such policy has allowed the technology importing domestic firms the independence of decision making so important for pursuing an autonomous path of expansion. The emergence of Japanese, and of late, Korean and Taiwanese corporations as multinational enterprises in their own right would not have been possible with technology acquisition under the FDI route. It must be emphasized, however, that market transfers alone do not guarantee such outcomes. They only provide necessary conditions and are effective only in the presence of responsive (and perhaps aggressive) local entrepreneurship willing to complement imported knowledge with extensive in-house technological effort on absorption, adaptation, continuous updating and eventually on innovation. In the absence of such effort, technology imports through markets may lead to, as the experience of India has shown, technological obsolescence. However, in view of the recent trends observed in Chapter 2 – revealing the reversal of the trend of licensing's increasing popularity as a mode of technology transfer since the mid-1980s – many technologies, especially the new core technologies, may not be available on arm's length basis any more. Yet arm's length licensing is still quite a popular mode of technology transfer in a number of sectors – especially those involving process technologies – and may allow certain possibilities of absorption and assimilation.

## *Effective competition policies are needed to regulate rent seeking by foreign and local monopolies*

The policy to support and protect national champions as proposed above, however, has a possible pitfall as it could lead to rent seeking by national champions. So the challenge for policy makers is to ensure that the national champions do not abuse their market power. Some competitive pressure is also necessary for them to sharpen their efficiency and to prompt them to pursue innovation based rivalry. The competition policy could take several forms depending upon the relative strength of local enterprises and the extent of monopoly. East Asian countries, for instance, have pushed competitiveness of national champions with intense domestic competition and by exposing them to international markets through rigorously monitoring imposed export targets.

The increasing domination of technologies by a few large corporations globally as observed in Chapter 2 coupled with technoprotectionism in industrialized countries and a recently strengthened and harmonized international regime of intellectual property protection also emphasize the need for effective anti-trust regulations in developing countries. The new norms of patent protection treat importing as working of patents and provide to the patent holders greater freedom to opt for exporting to particular markets and choosing not to transfer technology. Furthermore, the recent spurt of the cross-border merger and acquisition activity between large MNEs has adverse effects on the market concentration in many host countries where the affiliates of the merged enterprises operate as competitors. While most industrialized countries are equipped with effective anti-trust regulations to deal with possible abuse of monopoly power of patent holders and anti-competitive effects of international mergers, most developing countries either lack anti-trust legislation designed to deal with such situations or their implementation is poor (see Kumar, 1982, for an analysis of India). International initiatives to create codes of conduct for transfer of technology and on activities of MNEs have failed to evolve any effective norms. Adoption of effective competition policy instruments either at national levels or collectively at regional levels (*à la* European Union) to minimize the adverse impact of the possible resort to restrictive business practices by patent holders, is also a possible item on the policy agenda of developing countries.

Another aspect of competition policies is to provide a level playing field for local enterprises *vis-à-vis* subsidiaries of MNEs which enjoy access to their parent's brand and trade names besides a number of other intangible assets especially in the context of liberalization of national economies to MNEs. An affiliation with established global chains of companies having a wide range of products and services lends a formidable edge to MNE affiliates over the national firms in the host markets. To use their advantages most effectively MNE affiliates tend to adopt non-price modes of rivalry characterized by a heavy reliance on marketing and advertising to differentiate their products. These strategies raise barriers for the entry of new firms and are referred to as 'contrived entry barriers' in the industrial organization literture. The empirical studies have revealed that the advantages enjoyed by MNE affiliates in the form of access to brand names etc. either push host country local firms to serve the price competitive lower ends of the markets or to seek alliances with MNEs to buy access to their brand names (see Kumar and Siddharthan, 1997, ch. 7 for a review). The policy of promoting healthy competition between local firms and MNE affiliates could take the form of either offsetting the market power of MNE affiliation and foreign brands through fiscal measures or assisting national firms to build their own brands and technological capability. It is in view of this that some countries attempt to restrict the use of foreign brand names for domestic operations and instead encourage national enterprises to develop local brand names. In the case of pharmaceutical industry, for instance, some countries encourage the use of generic – rather than brand – names to curb the market power of MNEs' brand names.

### Policies for diffusion of knowledge

Another sphere where governmental intervention may be required to maximize gains from globalization is in diffusion of knowledge brought in by foreign enterprises. An important channel of diffusion of knowledge brought in by MNEs in the host economy is vertical inter-firm linkages with domestic enterprises. The trends with respect to local vertical inter-firm linkages has been examined in a couple of studies in this volume. We have observed rather slow progress in terms of generation of vertical linkages with local firms in developing host countries by affiliates of either US or Japanese MNEs. In addition, FDI abroad by Japanese MNEs tends to transplant the vendor-OEM links prevailing in the home country in that the traditional component suppliers of Japanese equipment manufacturers follow the latters' overseas investments. This pattern tends to limit the generation of local linkages. Therefore, policies promoting generation of vertical inter-firm linkages between MNE affiliates and local enterprises will help in diffusion of knowledge brought in by them. This knowledge diffusion could also be accomplished by creating sub-national or sub-regional clusters of inter-related activities which facilitate the spillovers of knowledge through informal and social contacts among the employees besides traditional buyer–seller links.

## International intervention in the area of international technology transfers and investments

The above findings also have implications for international intervention in the area of international business.

### FDI inflows are no substitute for concessional development finance

As observed earlier, there is a question mark on the ability of low income and least developed countries in attracting substantial magnitudes of FDI inflows, especially from the conventional Western sources, despite their recent expansion. This suggests that the expansion of FDI inflows, while a welcome development by itself, can hardly compensate for the shrinking levels of official development finance both in nominal as well as in real terms over the recent years which are generally expected to be directed to the poorer countries. The process of industrialization of least developed and low income countries cannot be left to market forces and FDI inflows alone. The international community will have to restore the flow of official resources and assistance to them to get the process of development started.

*An institutional framework for promoting FDI flows between developing countries could be fruitful*

As observed earlier, the efforts of low income and the least developed countries in attracting FDI inflows may be more successful if they targeted the flows originating in emerging sources – such as NIEs and other developing countries – than those originating in the industrialized countries, the more conventional sources. There appears to be scope for institutional intermediation at the regional or international level to direct these flows to the poorer countries, given the constraints of information and financing. On the part of receiving countries, a specific targeting of developing country FDI may be desirable. Some countries, such as Columbia and Costa Rica, have already begun to successfully target them. Developing country enterprises, especially in the early stages of their development, can be expected to have scarcity of funds to finance their overseas operations on a basis comparable to their Western counterparts. Hence, the institutional infrastructure for promoting inter-developing country FDI and technology transfers may include creation of a special fund for funding feasibility studies and for venture capital type financial support to projects based on developing country investments and technologies. In addition, some mechanisms for information dissemination among business enterprises of developing countries could be created. A networking of the chambers of commerce in developing countries could also be fruitful. Some initiatives of this type have already been taken at the regional level but need to be made more effective.

*Investment incentives offered by industrialized countries distort the pattern of FDI and hence need to be regulated*

A large number of industrialized country governments, especially the state, regional or local governments, extend large investment incentives to attract FDI inflows to particular regions. UNCTAD (1995: table VI.3) has listed some prominent examples of investment incentives granted by state or local governments in several industrialized countries. These include Setubal, Portugal offering $483.5 million incentive in 1991 to Auto Europa plant amounting on a per employee basis to $254,451, or Tuscaloosa, Albama in the US offering $250 million to the Mercedes-Benz plant in 1993 which works out to be a subsidy of $166,667 per worker employed. Urata has found the location of Japanese FDI to have been influenced by investment incentives. These incentives tend to distort the pattern of location of investments. Since resource scarce developing country governments can never match the incentives offered by the industrialized countries, they affect the investment climate in developing countries in relative terms or push them to indulge in costly incentive wars. Industrialized country governments – federal as well as local – need to be brought under a certain discipline to regulate investment incentives to ensure that the pattern of distribution of FDI inflows is not distorted.

*A multilateral regime on investment of the type being pushed by OECD countries will be inimical to developing countries' interests*

The OECD has launched negotiations in 1995 to establish a Multilateral Agreement on Investment (MAI) as a legally binding treaty open to even non-OECD member states to ensure higher standards of protection and legal security for foreign investors. In addition, the EU and Canada have also made a proposal to create a Possible Multilateral Framework on Investment (PMFI) under the auspices of the World Trading Organization (WTO) in 1996, with OECD's MAI providing a model (even if not to be adopted bodily). This has led to the setting up of a working group in WTO to consider the proposal. In any case OECD expects the treaty to become a sort of benchmark for investors to rate the treatment accorded to foreign investors. It will pressure developing countries desirous of attracting FDI to adopt it. The core of the OECD's draft MAI is based on the principles of national treatment and most-favoured-nation to foreign investors applicable to both pre- and post-investment phases. The implication of these provisions will be that the host countries will not be able to accord a more favourable treatment to local enterprises over foreign enterprises, although favourable treatment to the latter is not excluded. Since the provisions will apply to both pre- as well as post-investment phases, the screening mechanisms established by host countries to select FDI projects will not be possible. Therefore, the provisions of MAI will conflict with the policies recommended in the previous section, namely, protection to national champions and a selective treatment of FDI. Furthermore, the MAI draft has been widely criticized for being highly asymmetric in terms of not balancing the rights and responsibilities of corporations (see Kumar, 1996d; Panchamukhi, 1996, among others). In the words of David Chaytor, a British Member of Parliament, in a Statement made to the House of Commons on 23 July 1997, the MAI draft

> provides for an unprecedented shift in power from national Governments...pursuing a range of social, enviornmental and economic objectives, to transnational corporations with one objective only. It creates new rights for transnational corporations, but says nothing about their responsibilities. It leaves responsibilities with national Governments, but takes away their rights.

Therefore, MAI is approaching the issue from a narrow investor/home country perspective. The host country concerns and rights are not taken care of in the proposed formulations. In any case signing it could not guarantee greater FDI flows to the least developed countries; they are largely marginalized by FDI inflows at present despite very liberal policy regimes. It would also diminish the ability of developing countries to pursue selective policies with respect to FDI to channel them in accordance with their developmental goals. It does not address the issues concerning restrictive business practices and anti-trust implications of MNEs' operations.

It is difficult to understand the provocation for these initiatives on the part of industrialized countries. The FDI regimes are becoming increasingly lib-eral over the past decade and will continue to move in that direction in the coming years given the competition among countries for FDI inflows. The multilateral trade negotiations leading to the establishment of WTO have already included an agreement on eliminating the Trade Related Investment Measures (TRIMs) which limits the ability of host governments to regulate the FDI inflows. An international framework for settling investment disputes – the International Convention on Settlement of Investment Disputes (ICSID) – already exists under the auspices of the World Bank. The Multilat-eral Investment Guarantee Agency (MIGA) has also been launched to protect and insure overseas investments against political risks such as expropriation, blocked currency transfers, breach of contract, war, revolution and insurrec-tion. There have not been any glaring cases of disputes that could not be settled through the existing framework. The argument given in favour of MAI is that it will obviate the need for concluding numerous bilateral invest-ment treaties. Bilateral treaties are concluded between countries to deal with specific issues of concern between a pair of countries and are much easier to conclude. A multilateral framework may not be able to provide a general solution of all the issues of bilateral concerns. If there is any need at all for international intervention it is for enforcing certain norms of responsible cor-porate behaviour given the unprecedented power that MNEs now enjoy. Finally, FDI, like domestic investments, concerns development more than trade. Hence, WTO is not an appropriate forum to deal with investments. The link between FDI and trade is ambiguous and some FDI inflows, such as the market seeking type, actually substitute exports. Therefore, protection in the host countries acts as an inducement for FDI inflows, as noted earlier.

To sum up, a multilateral framework on investment either under OECD's auspices or a similar regime within WTO would not be in the interest of developing countries. The developing countries would do well to resist such attempts as they have so far done. Solidarity on the part of developing countries on an issue having an important bearing on their developmental prospects such as this would be important in projecting their concerns in a more effective manner in the international fora.

# BIBLIOGRAPHY

Agarwal, J. P. (1980) 'Determinants of Foreign Direct Investment: A Survey', *Weltwirtschaftliches Archiv* 116 (4): 739–73.

—— (1985) *Pros and Cons of Third World Multinationals: A Case Study of India*, Kieler Studien 195, Tübingen: J. C. B. Mohr.

—— (1994) 'The Effects of the Single Market Programme on Foreign Direct Investment into Developing Countries', *Transnational Corporations* 3 (2): 29–44.

—— (1996a) 'Impact of "Europe Agreements" on FDI in Developing Countries', *International Journal of Social Economics* 23 (10/11): 150–63.

—— (1996b) *European Union Direct Investment in ASEAN: Present Status, Future Direction and Policy Implications*. Paper presented at the Seminar on Promotion of Foreign Direct Investment in the Context of ASEAN Free Investment Area, 23–24 May, Bangkok.

Agarwal, J. P., Gubitz, A. and Nunnenkamp, P. (1991) *Foreign Direct Investment in Developing Countries. The Case of Germany*, Kieler Studien 238, Tübingen.

Agarwal, J. P., Hiemenz, U. and Nunnenkamp, P. (1995) *European Integration: A Threat to Foreign Investment in Developing Countries*, Kiel: Kiel Discussion Papers 246.

Aitken, B., and Harrison, A. (1993) 'Does Proximity to Foreign Firms Induce Technology Spillovers?'. PRD Working Paper, World Bank.

Aitken, B. Harrison, A. and Lipsey, R. E. (1996) 'Wages and Foreign Ownership: A Comparative Study of Mexico, Venezuela, and the United States', *Journal of International Economics* 40 (1/2).

Arita, T. and Fujita, M. (1996) *Local Agglomeration and Global Networks of the Semiconductor Industry: A Comparative Study of US and Japanese Firms*, University of Pennsylvania and Kyoto University (mimeo).

Audretsch, D. B. and Yamawaki, H. (1988) 'R&D Rivalry, Industrial Policy and US-Japanese Trade', *Review of Economics and Statistics* 70: 438–47.

Balasubramanyam, V. N. (1996) 'Software in South India', Discussion Paper, Department of Economics, Lancaster University.

Bank of Korea (1994) *Economic Statistics Yearbook, 1994*, Seoul.

—— (1995) *Overseas Direct Investment Statistics Yearbook*, Seoul, Bank of Korea: Foreign Exchange Department.

Barrell, R., Pain, N. and Hubert, F. (1996) *Regionalism, Innovation and the Location of German Direct Investment*, National Institute of Economic and Social Research, Discussion Paper 91, London.

216

Beifuβ, J. (1996) *Erfahrung deutscher Auslandsinvestoren in Reformländern Mittel-und Osteuropas*, Beiträge zur Wirtschafts-und Sozialpolitik 232, Institut der deutschen Wirtschaft, Cologne: Deutscher Institutsverlag.

Blomström, M. and Lipsey, R. E. (1993) 'Foreign Firms and Structural Adjustment in Latin America: Lessons from the Debt Crisis', in Hansson, G., ed., *Trade, Growth, and Development: The Role of Politics and Institutions*, London: Routledge.

Blomström, M., Lipsey, R. E. and Zejan, M. (1994) 'What Explains the Growth of Developing Countries', in Baumol, W. J., Nelson, R. R. and Wolff, E. N., eds, *Convergence of Productivity: Cross-National Studies and Historical Evidence*, New York: Oxford University Press: 243–59.

—— (1996) 'Is Fixed Investment the Key to Economic Growth?', *Quarterly Journal of Economics* 111 (1) February: 269–76.

Blomström, M. and Wolff, E. N. (1994) 'Multinational Corporations and Productivity Convergence in Mexico,' in Baumol, W. J., Nelson, R. N. and Wolff, E. N., eds, *Convergence of Productivity: Cross-National Studies and Historical Evidence*, New York: Oxford University Press: 263–84.

Buckley, P. J. and Casson, M. (1976) *The Future of the Multinational Enterprise*, London: Macmillan.

Cantwell, J. A. and Hodson, C. (1991) 'Global R&D and British Competitiveness', in Casson, M. C., ed., *Global Research Strategy and International Competitiveness*, Oxford: Basil Blackwell: 133–82.

Casson, M. C., ed. (1991) *Global Research Strategy and International Competitiveness*, Oxford: Basil Blackwell.

Caves, R. E. (1971) 'International Corporations: The Industrial Economics of Foreign Investment', *Economica* 38: 1–27.

—— (1996) *Multinational Enterprise and Economic Analysis*, 2nd edn, Cambridge: Cambridge University Press.

Chaponniere, J. R. (1992) 'The Newly Industrialising Economies of Asia, International Investment and Transfer of Technology', *STI Review* 9, Paris: OECD.

China, Republic of (1995) *Statistical Yearbook of the Republic of China*, Taipei, Directorate General of Budget, Accounting, and Statistics, November.

—— (1996) *Quarterly National Economic Trends, Taiwan Area, The Republic of China*. Taipei, Directorate General of Budget, Accounting, and Statistics, May.

Contractor, F. J. (1990) 'Do Government Policies toward Foreign Investment Matter?', GSM Working Paper no. 90–15, Newark, NJ: Rutgers University.

Correa, C. M. (1997) 'New International Standards for Intellectual Property: Impact on Technology Flows and Innovation in Developing Countries', *Science and Public Policy* 24(2): 79–92.

Dahlman, C., ul Haque, I. and Takeuchi, K. (1995) 'The World Trading Environment', in ul Haque, I. *et al.*, *Trade, Technologies, and International Competitiveness*, Washington DC: The World Bank: 155–78.

Dasgupta, S., Mody, A. and Sinha, S. (1995) *Japanese Multinationals in Asia: Capabilities and Motivation*, Washington: World Bank (mimeo).

Davidson, W. H. and McFetridge, D. G. (1985) 'Key Characteristics in the Choice of International Technology Transfer', *Journal of International Business Studies* 16 (Summer): 5–21.

217

Deardoff, A. V. (1984) *Comparative Advantages and International Trade and Investment in Services*. Seminar Discussion Paper 137, Department of Economics, University of Michigan.

Deutsche Bundesbank (1997) 'Die Aktie als Finanzierungs-und Anlageinvestment', *Monatsbericht*, January: 28–41, Frankfurt/Main.

—— (various issues). *Kapitalverflechtung mit dem Ausland*. Statistische Sonderveröffentlichung 10, Frankfurt/Main.

Deutscher Industrie-und Handelstag (DIHT) (1996) *Produktionsverlagerung als Element der Auslandsinvestitionen. Ergebnisse einer Unternehmensbefragung im Herbst 1996*, Bonn.

Deutsches Institut für Wirtschaftsforschung (DIW) (1995) 'Passive Lohnveredelung im Rahmen der Textil-und Bekleidungsimporte Deutschlands und der EU', *Wochenbericht* 62 (17): 338–46, Berlin.

Dicken, P. (1994) 'Global–Local tensions: Firms and States in a Global Space-economy', *Economic Geography* 70 (2): 101–28.

Dunning, J. H. (1981) *International Production and Multinational Enterprise*, London: Allen & Unwin.

—— (1985) 'The United Kingdom', in Dunning, J. H., ed., *Multinational Enterprises, Economic Structure and International Competitiveness*, Chichester and New York: John Wiley and Sons.

—— (1990) *The Globalization of Firms and the Competitiveness of Nations*, Lund (Sweden): University of Lund, The Crafoord Lectures 1989 : 9–57.

—— (1993) *Multinational Enterprises and the Global Economy*, Wokingham, UK: Addison Wesley.

—— (1994a) *Globalization, Economic Restructuring and Development*, Geneva: UNCTAD, the 6th Prebisch Lecture.

—— (1994b) 'Multinational Enterprises and the Globalization of Innovatory Capacity', *Research Policy* 23: 67–88.

—— (1995a) 'Reappraising the Eclectic Paradigm in the Age of Alliance Capitalism', *Journal of International Business Studies* 26 (3): 461–91.

—— (1995b) 'What's Wrong – and Right – with Trade Theory?', *International Trade Journal* 9 (2): 153–202.

—— (1996a) 'Explaining Foreign Direct Investment in Japan: Some Theoretical Insights', in Yoshitomi, M. and Graham, E., eds, *Foreign Direct Investment in Japan*, London: Edward Elgar: 8–63.

—— (1996b) 'The Geographical Sources of the Competitiveness of Firms: Some Results of a New Survey', *Transnational Corporations* 5 (3): 1–30.

—— (1997a) 'The European Internal Market Program and Inbound Foreign Direct Investment' (in two parts), *Journal of Common Market Studies* 35 (1 and 2): 1–30, and 189–224 respectively.

—— (1997b) 'Does Ownership Really Matter in a Globalizing Economy?', in Woodward, D. P. and Nigh, D. W., *Beyond Us and Them: Foreign Ownership and US Competitiveness in the 1990s*, Westport, CE: Quorum Books.

Dunning, J. H. and Narula, R. (1995) 'The R&D Activities of Foreign Firms in the US', *International Studies of Management and Organization* 25 (1–2) Spring–Summer: 39–75.

Dunning, J. H. and Narula, R. (eds) (1996) *Foreign Direct Investment and Governments*, London and New York: Routledge.

Dunning, J. H., van Hoesel, R. and Narula, R. (1998) 'Explaining the New Wave of Outward FDI from Developing Countries', *International Business Review* 7(6).

Economic Planning Agency (1997) *The Spread of Globalization and Issues for the 21st Century*. Report of the 21st Century Committee, Economic Council, the Government of Japan.

Enright, M. J. (1994) 'Regional Clusters and Firm Strategy'. Paper presented to Prince Bertil Symposium, *The Dynamic Firm, The Role of Regions, Technology, Strategy and Organization*, Stockholm, June.

European Commission and UNCTAD Division on Transnational Corporations and Investment (EC–UNCTAD) (1996) *Investing in Asia's Dynamism: European Union Direct Investment in Asia*, Luxembourg.

EUROSTAT (1995) *European Union Direct Investment 1984–93*, European Commission, Luxembourg: Office for Official Publications of the European Communities.

Findlay, R. (1978) 'Relative Backwardness, Direct Foreign Investment, and the Transfer of Technology: A Simple Dynamic Model', *Quarterly Journal of Economics* 92(1) February: 1–16.

Florida, R. (1995) 'Towards the Learning Region', *Futures* 27 (5): 527–36.

Freeman, C. and Hagedoorn, J. (1992) 'Globalization of Technology', *A Report for the FAST Programme*, Commission of the European Communities. Theme C: Global Perspective 2010 – Tasks for Science and Technology, vol. 3–4. Brussels: EC.

Gang, Y. (1992) 'Chinese Transnational Corporations', *Transnational Corporations* 1 (2): 125–33.

Granstrand, O., Hakanson, L. and Sjolander, S., eds (1992) *Technology, Management and International Business: Internationalization of R&D and Technology*, Chichester, UK: Wiley.

Gray, A., Golog, E. and Markusen, A. (1995) *Big Firms, Long Arms, Wide Shoulders: The Hub and Spoke Industrial District in the Seattle*, New Brunswick, NJ: PRIE Working Paper no. 79.

Guisinger, S. E. and Associates (1985) *Investment Incentives and Performance Requirements*, New York: Praeger.

Haddad, M. and Harrison, A. (1993) 'Are There Positive Spillovers from Foreign Direct Investment? Evidence from Panel Data for Morocco', *Journal of Development Economics*, October.

Hagedoorn, J. (1993) 'Strategic Technology Alliances and Modes of Cooperation in High Technology Industries', in Grabher, G., ed., *The Embedded Firm*, London and Boston: Routledge: 116–37.

—— (1996) 'Trends and Patterns in Strategic Technology Partnering Since the Early Seventies', *Review of Industrial Organization* 11: 601–16.

Han, S.-T. (1992) *European Integration: The Impact on Asian Newly Industrialising Economies*, Development Centre Documents, Paris: OECD.

Harrison, A. (1996) 'Determinants and Effects of Direct Foreign Investment in Côte d'Ivoire, Morocco and Venezuela', in Roberts, M. and Tybout, J., eds, *Industrial Evolution in Developing Countries*, Oxford: The World Bank and Oxford University Press: 163–86.

Harrison, B. (1992) 'Industrial Districts: Old Wine in New Bottles', *Regional Studies* 26 (5): 469–83.

—— (1994) *Lean and Mean: The Changing Landscape of Corporate Power in the Age of Flexibility*, New York: Basic Books.

Hennart, J.-F. and Park, Y. R. (1994) 'Location, Governance and Strategic Determinants of Japanese Manufacturing Investment in the US', *Strategic Management Journal* 15: 419–36.

Hiemenz, U. *et al.* (1994) *Regional Integration in Europe and its Effects on Developing Countries*, Kieler Studien 260, Tübingen.

Hirsch, S. (1970) 'Technological Factors in the Composition and Direction of Israel's Industrial Exports', in Vernon, R., ed., *The Technology Factor in International Trade*, New York: NBER: 145–231.

Hoey, J. (1996) *Germany's New Drive Eastwards. Economies in Transition, Eastern Europe and the Former Soviet Union, Regional Overview (EIU)*, 1st Quarter: 5–16.

Hymer, S. H. (1960) (published 1976) *The International Operations of National Firms: A Study of Direct Foreign Investment*, Cambridge, MA: MIT Press.

IADB-IRELA (1996) *Foreign Direct Investment in Latin America in the 1990s*, Madrid: Inter-American Development Bank and Institute for European-Latin American Relations.

IMF (various years) *Balance of Payments Yearbook* (summary tables), Washington: IMF.

—— (1995) *International Financial Statistics Yearbook*, Washington, DC: International Monetary Fund.

Jeon, Y.-D. (1992) 'The Determinants of Korean Foreign Direct Investment in Manufacturing Industries', *Weltwirtschaftliches Archiv* 128 (3): 526–43.

JETRO (1994) *White Paper on Foreign Direct Investment, 1994, Japanese Corporations Search for New Approaches*, Tokyo: Japan External Trade Organization.

Joekes, S. (1982) 'The Multifibre Arrangement and Outward Processing: the Case of Morocco and Tunisia', in Stevens, C. and Themat, J. V., eds, *EEC and the Third World: A Survey*, London and Sydney: Hodder & Stoughton: 102–12.

Kawai, M. and Urata, S. (1995) 'Are Trade and Direct Investment Substitutes or Complements? An Analysis of the Japanese Manufacturing Industry', Paper presented at The International Conference on Economic Development and Cooperation in the Pacific Basin, Berkeley, CA, June.

Kenney, M. and Florida, R. (1993) 'The Organization and Geography of Japanese R&D: Results from a Survey of Japanese Electronics and Biotechnology Firms', *Research Policy* 23: 305–23.

—— (1994) 'The Globalization of Japanese R&D: The Economic Geography of Japanese R&D Investment in the US', *Economic Geography* 70: 344–69.

Kim, K. D. and Sang, H. L. (1990) 'The Role of the Korean Government in Technology Import', in Lee, C. H. and Yamazawa, I., eds, *The Economic Development of Japan and Korea: A Parallel with Lessons*, New York: Praeger.

Kim, L. (1997) *Imitation to Innovation: The Dynamics of Korea's Technological Learning*, Boston, MA: HBS Press.

Klodt, H. and Maurer, R. (1996) *Internationale Direktinvestitionen: Determinanten und Konsequenzen für den Standort Deutschland, Discussion Paper 284, Kiel: Institute of World Economics*.

Kogut, B. (1983) 'Foreign Direct Investment as a Sequential Process', in Kindleberger, C. P. and Audretsch, D. B., eds, *The Multinational Corporation in the 1980s*, Cambridge, MA: MIT Press.

Kravis, I. B. and Lipsey, R. E. (1982) 'The Location of Overseas Production and Production for Export by US Multinational Firms', *Journal of International Economics* 12: 201–23.

Krugman, P. (1990) *Rethinking International Trade*, Cambridge, MA: MIT Press.

Kuemmerle, W. (1996) *The Drivers of Foreign Direct Investment into Research and Development: An Empirical Investigation*, Boston: Harvard Business School Working Paper no. 96:062.

Kumar, N. (1982) 'Regulating Multinational Monopolies in India', *Economic and Political Weekly* 17 May, 909–18.

—— (1986) 'Foreign Direct Investments and Technology Transfers among Developing Countries', in Panchamukhi, V. R. *et al.*, *The Third World and the World Economic System*, New Delhi: Radiant: 139–65.

—— (1987) 'Intangible Assets, Internationalisation and Foreign Production: Direct Investments and Licensing in Indian Manufacturing', *Weltwirtschaftliches Archiv* 123: 325–45.

—— (1990) 'Internalisation of Technology Transfer by U.S. Multinationals: A Transactions Cost Perspective', Paper presented at the Annual Meeting of the Academy of International Business, Toronto, October.

—— (1994a) 'Determinants of Export Orientation of Foreign Production by U.S. Multinationals: An Inter-country Analysis', *Journal of International Business Studies* 25(1): 141–56.

—— (1994b) 'Regional Trading Blocs, Industrial Reorganizations and Foreign Direct Investments – The Case of the Single European Market', *World Competition* 18(2): 35–55.

—— (1994c) *Multinational Enterprises and Industrial Organization: The Case of India*, New Delhi, Thousand Oaks and London: Sage Publications.

—— (1995) 'International Linkages, Technology and Exports of Developing Countries: Trends and Policy Implications'. INTECH Discussion Papers no. 9507, Institute of New Technologies, The United Nations University, The Netherlands.

—— (1996a) 'Multinational Enterprises, New Technologies and Export-Oriented Industrialization in Developing Countries: Trends and Prospects'. INTECH Discussion Papers no. 9602, Institute of New Technologies, The United Nations University, The Netherlands.

—— (1996b) 'Intellectual Property Protection, Market Orientation and Location of Overseas R&D Activities by Multinational Enterprises', *World Development* 24(4): 673–88.

—— (1996c) 'India: Industrialization, Liberalization and Inward and Outward Foreign Direct Investment', in Dunning, J. H. and Narula, R., eds, *Foreign Direct Investment and Governments*, London: Routledge: 348–79.

—— (1996d) 'A Multilateral Regime on Investments: A Note on Understanding the Developing Country's Concerns', Paper presented at the International Workshop on FDI, Technology Transfer and Export-Orientation, UNU/INTECH, Maastricht, 15–16 November 1996.

—— (forthcoming) *Globalization and Quality of Foreign Direct Investment: A Quantitative Explanation of the Role of Multinationals in Host Country Industrialization, Export Expansion and Innovation*, London and New York: Routledge and UNU Press.

Kumar, N. and Siddharthan, N. S. (1997) *Technology, Market Structure and Internationalization: Issues and Policies for Developing Countries*, London and New York: Routledge.

Lall, S. (1980) 'International Technology Market and Developing Countries', *Economic and Political Weekly*, Annual Number, February: 311–32.

—— (1984) 'Exports of Technology by Newly-industrializing Countries: An Overview', *World Development* 12 (5/6): 471–80.

—— (1997) 'Technological Change and Industrialization in the Asian NIEs: Achievements and Challenges', in Nelson R. R. and Kim L., eds, *Industrialization and Competitiveness in Newly Industrializing Economics*.

Lall, S. *et al.* (1983) *The New Multinationals: The Spread of Third World Enterprises*, Chicester: John Wiley.

Lecraw, D. J. (1977) 'Direct Investment by Firms from Less Developed Countries', *Oxford Economic Papers* 29: 442–57.

Lewis, C. (1938) *America's Stake in International Investments*, Washington, DC: The Brookings Institution.

Lipsey, R. E. (1995) 'Trade and Production Networks of U.S. MNCs and Exports by their Asian Affiliates', Cambridge, MA: NBER Working Paper no. 5255, September.

Lipsey, R. E., Blomström, M. and Kravis, I. B. (1990) 'R&D by Multinational Firms and Host Country Exports', in Evenson, R. E. and Ranis, G., eds, *Science and Technology: Lessons for Development Policy*, Boulder, and San Francisco, CA: Westview Press: 271–300.

Lipsey, R. E., Blomström, M. and Ramstetter, E. (1995) 'Internationalized Production in World Output', Cambridge, MA: NBER Working Paper no. 5385, December.

Lipsey, R. E. and Weiss, M. Y. (1981) 'Foreign Production and Exports in Manufacturing Industries', *Review of Economics and Statistics* 63 (4): 488–94.

Loree, D. W. and Guisinger, S. E. (1995) 'Policy and Non-Policy Determinants of U.S. Equity Foreign Direct Investment', *Journal of International Business Studies* 26(2): 281–300.

Luttmer, E. and Oks, D. (1993) 'Productivity in Mexican Manufacturing Industries', Washington, DC: World Bank.

MacCormack, A. D., Newman III, L. J. and Rosenfield, D. B. (1994) 'The New Dynamics of Global Manufacturing Site Location', *Sloan Management Review* 35, Summer: 69–80.

MacDougall, G. D. A. (1960) 'The Benefits and Costs of Private Investment from Abroad: A Theoretical Approach', *Economic Record* March: 13–35.

Madeuf, B. (1984) 'International Technology Transfers and International Technology Payments: Definitions, Measurements and Firms' Behaviour', *Research Policy* 13: 125–40.

Markusen, J. R. (1995) 'The Boundaries of Multinational Enterprises and the Theory of International Trade', *Journal of Economic Perspectives* 9 (2): 169–89.

Marshall, A. (1920) *Principles of Economics*, 8th edn, London: Macmillan.

Mataloni, R. J. Jr (1995) 'U.S. Multinational Companies: Operations in 1993', *Survey of Current Business* 75 (6) June: 31–51.

—— (1997) 'U. S. Multinational Companies: Operations in 1995', *Survey of Current Business* vol. 77, no. 10, October, 44–68.

Mataloni, R. J. Jr and Fahim-Nadar, M. (1996) 'Operations of U.S. Multinational Companies: Preliminary Results from the 1994 Benchmark Survey', *Survey of Current Business* vol. 76, no. 12, December, 11–37.

Mataloni, Raymond, J., Jr and Goldberg, L. (1994) 'Gross Product of U.S. Multinational Corporations, 1977–91', *Survey of Current Business* vol. 74, no. 2, February 42–63.

Nunnenkamp, P., Gundlach, E. and Agarwal, J. P. (1994) *Globalisation of Production: Consequences for National Trade Policies*, Kieler Studien 262, Tübingen.

OECD (1982) *North/South Technology Transfer: The Adjustments Ahead*, Paris: OECD.

—— (1995a) *Industry and Technology: Scoreboard of Indicators*, Paris: OECD.

—— (1995b) *International Direct Investment Statistics Year Book 1995*, Paris: OECD.

—— (1996) *International Direct Investment Statistics Yearbook 1996*, Paris: OECD.

Ohmae, K. (1995) *The End of the Nation State: The Rise of Regional Economies*, London: Harper.

Okamoto, Y. and Urata S. (1994) 'Japan's Foreign Direct Investment in Malaysia in the 1990s,' Paper presented at the Third Annual Conference on Japan, 'Revitalization of Japan's Economy: Implications for Malaysia', Kuala Lumpur, April.

Ozawa, T. (1996) 'Japan: The Macro-IDP, Meso-IDPs and the Technology Development Path (TDP)', in Dunning, J. H. and Narula, R., eds, *Foreign Direct Investment and Governments*, London and New York: Routledge: 142–73.

Pack, H. and Westphal, L. E. (1986) 'Industrial Strategy and Technological Change: Theory versus Reality', *Journal of Development Economics* 22: 87–128.

Panchamukhi, V. R. (1996) *Multilateral Agreement on Investment (MAI): What Should be the Response of the Developing Countries*, Geneva: The South Centre.

Park, S. O. and Markusen, A. R. (1992) *New Industrial Districts: A Critique and Extension from the Developing World*, Paper presented at the Symposium of the IGU Commission on Industrial Change, Time, Space, Competition and Contemporary Industrial Change, Florida, August.

—— (1995) 'Generalizing New Industrial Districts: A Theoretical Agenda and an Application from a non-Western Economy', *Environment and Planning* A 27: 81–104.

Pearce, R. D. and Singh, S. (1992) 'Internationalization of R&D among the World's Leading Enterprises', in Granstrand, O., Håkanson, L. and Sjolander, S., eds, *Technology, Management and International Business: Internationalization of R&D and Technology*, Chichester (UK): Wiley.

Perez, C. (1983) 'Structural Change and Assimilation of Technologies in the Economic and Social Systems', *Futures* 15, October: 357–75.

Peres, N. W. (1993) 'The Internationalization of Latin American Industrial Firms', *Cepal Review* 49: 55–74.

Petri, P. A. (1995) 'The Interdependence of Trade and Investment in the Pacific', in Chen, E. K. Y. and Drysdale, P. eds, *Corporate Links and Foreign Direct Investment in Asia and the Pacific*, NSW, Australia: Harper Educational Publishers.

Pomfret, R. (1986) *Mediterranean Policy of the European Community*, London: Macmillan.

Porter, M. E. (1990) *The Competitive Advantage of Nations*, New York: Free Press: 42–63.

Reddy, M. N. and Zhao, L. (1990) 'International Technology Transfer: A Review', *Research Policy* 19: 285–307.

Romer, P. (1993) 'Idea Gaps and Object Gaps in Economic Development', *Journal of Monetary Economics* 32 (3), December: 543–73.

Rosenberg, N. and Frischtak, C., eds (1985) *International Technology Transfer*, New York: Praeger.

Rugman, A. M. and D'Cruz, J. R. (1995) *The Five Partners Business Network Model*, Paper presented to the conference on The Multinational Enterprise in the 21st Century, Taipei: Chinese Culture University, November.

Ruigrok, W. and Val Tulder, R. (1995) *The Logic of International Restructuring*, London and New York: Routledge.

Saxenian, A. L. (1994) *Regional Advantage: Culture and Competition in Silicon Valley and Route 128*, Cambridge, MA: Harvard University Press.

Scherer, F. M. and Belous, R. S. (1994) *Unfinished Tasks: The New International Trade Theory and the Post-Uruguay Round Challenges*, London and Washington: British-North American Committee.

Schoenberger, E. (1988) 'Multinational Corporations and the New International Division of Labor: A Critical Appraisal', *International Regional Science Review* 11 (2): 105–19.

Stewart, F. (1979) *International Technology Transfer: Issues and Policy Options*, Washington DC: World Bank (Staff Working Paper no. 344).

Stopford, J. M. (1992) *The World Directory of Multinational Enterprises*, Basingstoke: Macmillan

Stopford, J. M., Dunning, J. H. and Haberich, K. O. (1980) *The World Directory of Multinational Enterprises, Basingstoke*: Macmillan

Storper, M. and Scott, A. J. (1995) 'Market Forces and Policy Imperatives in Local and Global Context', *Futures* 27 (5): 505–26.

Suzuki, Y. (1994) *Multinationalization of Japanese Manufacturing: International Locational Practices*, University of Reading Department of Economics, Discussion Papers in International Investment and Business Studies, vol. V, no. 183, May.

Swamidaas, P. M. (1990) 'A comparison of the Plant Location Strategies of Foreign and Domestic Manufacturers in the US', *Journal of International Business Studies* 21 (2): 301–18.

Taiwan, MOEA (1993) *Outward Investment From Taiwan ROC, A New Source of International Capital*, Taipei: Ministry of Economic Affairs, Industrial Development and Investment Centre.

Tho, V. T. (1993) 'Technology Transfer in the Asian Pacific Region: Implications of Trends since the Mid-1980s', in Ito, T. and Krueger, A., eds (1993) *Trade and Protectionism*, Chicago and London: University of Chicago Press (NBER): 243–67.

Tolentino, P. E. E. (1993) *Technological Innovation and Third World Multinationals*, London and New York: Routledge.

Tuldar, R. van and Junne, G. (1988) *European Multinationals in Core Technologies*, London: Wiley.

Tyler, W. G. (1981) 'Growth and Export Expansion in Developing Countries, Some Empirical Evidence', *Journal of Development Economics* 9: 121–30.

Ulgado, F. and Yu Chwo-Ming, J. (1994) *Multinational Enterprises from Developing Countries: Location Decision Characteristics of FDI in the US*, Atlantic Georgia Tech CIBER Working Paper 94–109, October.

UNCTAD (1993a) *World Investment Report 1993: Transnational Corporations and Integrated Production*, New York and Geneva: UN.

——(1993b) *World Investment Directory. Vol III, Developed Countries*, New York: UN.

——(1994a) *World Investment Report 1994: Transnational Corporations, Employment and the Workplace*, New York and Geneva: UN.

——(1994b) *World Investment Directory, Foreign Direct Investment, Legal Framework and Corporate Data. Vol. IV: Latin America and the Caribbean*, New York: Division on Transnational Corporations and Investment.

—— (1995a) *World Investment Report 1995. Transnational Corporations and Competitiveness*, Geneva.

—— (1995b) *Incentives and Foreign Direct Investment*, Geneva: Background report of Commission on International Investment and Transnational Corporations, April 1995 (TD/B/ITNC/Misc 1).

—— (1995c) Recent Developments in International Investment and Transnational Corporations, Trends in Foreign Direct Investment, TD/B/ITNC/2, Geneva: United Nations Conference on Trade and Development.

—— (1996a) *World Investment Report 1996: Investment, Trade and International Policy Arrangements*, New York and Geneva.

—— (1996b) *Sharing Asia's Dynamism: Asian Direct Investment in the European Union*, New York and Geneva: UN.

—— (1997) *World Investment Report 1997*, Geneva: United Nations Conference on Trade and Development.

UNCTC (1988) *Transnational Corporations and World Development*, New York, UN.

—— (1992) *The Determinants of Foreign Direct Investment: A Survey of the Evidence*, New York: United Nations Centre on Transnational Corporations.

UNDP (1992) *Human Development Report 1992*, New York: UN.

UNESCO (1996) *World Science Report 1996*, Paris: UNESCO.

United Nations (UN) (1992) *World Investment Directory 1992. Foreign Direct Investment Framework and Corporate Data. Vol. II, Central and Eastern Europe*, New York: Transnational Corporations and Management Division and Economic Commission for Europe.

—— (1993) *MSPA Handbook of World Development Statistics*, Department of Economic and Social Information and Policy Analysis, New York: June.

—— (various issues) *East-West Investment News*, Geneva: Economic Commission for Europe.

UN-TCMD (1993) *Transnational Corporations from Developing Countries, Impact on Their Home Countries*, New York: United Nations, Transnational Corporations and Management Division, Department of Economics and Social Development.

Urata, S. (1990) 'The Impact of Imported Technologies on Japan's Economic Development', in Lee, C. H. and Yamazawa, I., eds, *The Economic Development of Japan and Korea: A Parallel with Lessons*, New York: Praeger: 73–86.

—— (1993a) 'Japanese Foreign Direct Investment and its Effect on Foreign Trade in Asia', in Ito, T. and Krueger, A. O., eds, *Trade and Protectionism*, Chicago: University of Chicago Press: 273–99.

—— (1993b) 'Obstacles to Further Economic Growth in East Asia and Japan's Economic Assistance', *Japan Review of International Affairs* 7(4), Fall: 297–315.

—— (1996) 'Japanese Foreign Direct Investment and Technology Transfer in Asia', Discussion Paper Series 4, APEC Study Center, Waseda University, Tokyo.

Urata, S. and Iriyama, A. (1997) 'Foreign Direct Investment and Technology Transfer in China', Japan Center for Economic Research, Tokyo, mimeo.

US Department of Commerce (various dates) 'US Business Enterprises Acquired or Established by Foreign Direct Investors', *Survey of Current Business*, usually in May issue.

—— (1981) *U.S. Direct Investment Abroad, 1977*, Washington, DC: Bureau of Economic Analysis.

——(1985) *U.S. Direct Investment Abroad: 1982 Benchmark Survey Data*, Washington, DC: Bureau of Economic Analysis.

——(1992) *U.S. Direct Investment Abroad: 1989 Benchmark Survey, Final Results*, Washington, DC: Bureau of Economic Analysis, US Government Printing Office, October.

——(1994) *U.S. Direct Investment Abroad: Operations of U.S. Parent Companies and their Foreign Affiliates, Revised 1991 Estimates*, Washington, DC: Bureau of Economic Analysis, June.

——(1995a) *U.S. Direct Investment Abroad: Operations of U.S. Parent Companies and their Foreign Affiliates, Revised 1992 Estimates*, Washington, DC: Bureau of Economic Analysis, June.

——(1995b) *U.S. Direct Investment Abroad: Operations of U.S. Parent Companies and their Foreign Affiliates, Preliminary 1993 Estimates*, Washington, DC: Bureau of Economic Analysis, June.

——(1996) 'U.S. Direct Investment Abroad: Detail for Historical-Cost Position and Related Capital and Income Flows, 1995', *Survey of Current Business* 76 (9): 98–128.

——(1997a) *U.S. Direct Investment Abroad: 1994 Benchmark Survey, Preliminary Results*, Bureau of Economic Analysis, Washington, DC: January.

——(1997b) 'Annual Revision of the National Income and Product Accounts: Annual Estimates, 1993–96, and Quarterly Estimates, 1931–1971', *Survey of Current Business*, vol. 77, no. 8, August, 6–167.

van Hoesel, R. (1996) 'Taiwan: Foreign Direct Investment and the Transformation of the Economy', in Dunning, J. and Narula, R., eds, *Foreign Direct Investment and Governments*, London: Routledge: 280–315.

Vernon, R. (1966) 'International Investment and International Trade in the Product Cycle', *Quarterly Journal of Economics* 80: 190–207.

Veugelers, R. (1991) 'Locational Determinants and Ranking of Host Countries: An Empirical Assessment', *Kyklos* 44 (3): 363–82.

Vickery, G. (1986) 'International Flows of Technology – Recent Trends and Developments', *STI Review* 1, Autumn, Paris: OECD.

——(1988) 'A Survey of International Technology Licensing', *STI Review,* December, Paris: OECD.

Wells, L. T., Jr (1983) *Third World Multinationals: The Rise of Foreign Investment from Developing Countries*, Cambridge, MA: MIT Press.

——(1993) 'Mobile Exporters: New Foreign Investors in East Asia', in Froot, K. A., ed., *Foreign Direct Investment*, Chicago: University of Chicago Press and NBER, 173–91.

Wheeler, D. and Mody, A. (1992) 'International Investment, Location Decisions: The Case of U.S. Firms', *Journal of International Economics* 33: 57–76.

World Bank (1979) *International Technology Transfer: Issues and Policy Options*, World Bank Staff Working Paper no. 344, Washington, DC: World Bank.

——(1989) *Foreign Direct Investment from Newly Industrialized Economies*, Washington DC: World Bank, Industry Development Division, Industry and Energy Department.

——(1993) *The East Asian Miracle: Economic Growth and Public Policy*, Oxford and New York: Oxford University Press.

——(1995) *World Tables: 1995*, Washington, DC: World Bank.

—— (1996) *World Development Report*, Oxford and New York: Oxford University Press.

Yuskavage, R. E. (1996) 'Improved Estimates of Gross Product by Industry, 1959–94', *Survey of Current Business* vol. 76, no. 8, August, 133–55.

Zhang, H. and Van Den Bulcke, D. (1996) 'China: Rapid Changes in the Investment Development Path', in Dunning, J. and Narula, R., eds, *Foreign Direct Investment and Governments*, London: Routledge: 380–422.

# INDEX